FUNDAMENTALS OF NUMERICAL CONTROL

William W. Luggen

DELMAR PUBLISHERS INC.

A very special thanks to my wife, Linda, for her patience and time spent helping to produce this text. It is to her and my beautiful daughters that I dedicate this book.

Most of all, I thank God for the opportunities He has given me.

Delmar staff

Administrative editor: Mark W. Huth
Project editor: Marjorie A. Bruce
Production editor: Ruth Saur

For information, address Delmar Publishers, Inc.
2 Computer Drive West, Box 15-015
Albany, New York 12212

Printed in the United States of America
Published simultaneously in Canada
by Nelson Canada,
A division of International Thomson Limited

10 9 8 7 6 5 4 3

Library of Congress Cataloging in Publication Data

Luggen, William W., 1947–
    Fundamentals of numerical control.

    Includes index.
    1. Machine-tools — Numerical control. I. Title.
TJ1189.L84   1983   621.9'023   83-71970
ISBN 0-8273-2162-7

CONTENTS

PREFACE

Since numerically controlled machine tools first began making their way into industry around 1957, they have increased considerably in control and machine tool capabilities. Related technological skills have increased as well. As we look into the future of manufacturing, it is apparent that it holds many opportunities for individuals who possess the process-type skills of manufacturing in general and, more specifically, as applied to numerical control and its manufacturing applications.

As manager of N/C programming, tool design, and manufacturing engineering systems for Cincinnati Milacron Inc. and teacher of N/C programming, William W. Luggen is aware of the needs of students in this field. The author's purpose in writing this textbook is to provide students with a fundamental yet detailed knowledge of numerical control from a conceptual point of view.

Every aspect of this text, from content to format, was designed with this one goal in mind — to lead students to a more thorough understanding of the subject matter as easily as possible. Special features are purposely incorporated into the text to help students attain this goal.

- *Subjective review questions* to promote original thinking, rather than mere memorization.

- An appendix of *useful formulas and tables* to use while studying the chapters to which they pertain.

- A *resource list of films* to supplement chapter study.

- *Detailed glossary* for definition of technical and key terms used in the text.

- *Objective questions* in the Instructor's Guide with answers for testing purposes.

- *Suggested activities* in the Instructor's Guide to aid in the students' understanding of chapter material.

- Appendix of *preparatory functions, miscellaneous functions,* and *other address characters* for quick and easy reference.

- List of *safety rules* to follow when using N/C machines.

As inflation continues to spiral, taking its toll on incomes and industrial and educational budgets, it has become virtually impossible for schools to afford modern numerically controlled machine tools. In addition, inflation has outpriced schools' capabilities to provide adequate tooling and maintenance procedures. Students desiring a knowledge of numerical control should not be denied the opportunity to learn its principles and concepts simply because the sophisticated N/C machines are unaffordable.

The first three chapters provide the student with a history of numerical control, its real purpose, and how it basically works. Also included are concepts of basic coding systems, axis notation, tapes, and tape preparation equipment. These chapters serve as a base for the rest of the text.

Subsequent chapters emphasize concepts and fundamentals, first in a general format and then in more specific formats of turning and machining centers.

The coverage of numerical control includes a discussion of computer applications to numerical control, specifically in the areas of APT programming. Tooling is discussed as well, primarily because it is the most underrated aspect of numerical control programming. Specific details of setup and fixtures have been purposely omitted; they are a study in themselves and will invariably differ from one manufacturer to another. Extensive calculations have not been included so as to concentrate more thoroughly on N/C programming principles and concepts.

ACKNOWLEDGMENTS

The author extends his sincere appreciation to Cincinnati Milacron Inc. for some of the essential materials used throughout this text and also for providing the atmosphere and opportunities to develop his talents. Special thanks also are expressed to Jack Cahall for providing the support needed to undertake and complete this project.

The author and publisher recognize with appreciation the technical assistance and photographs which the following companies provided:

Bridgeport Machines
Cincinnati Milacron Inc.
Cleveland Twist Drill
DoAll Company
Industrial Controls Division, Bendix Corporation
Kearney & Trecker Corporation
Kennametal, Inc.
Lodge & Shipley Company
Monarch Machine Tool Company
Morse Cutting Tools Division, Gulf & Western Manufacturing Company
Portage Machine Company
Sharpaloy Division, Precision Industries, Inc., Centerdale, R.I.
The Valeron Corporation
Turning Machine Division, The Warner & Swasey Co., subsidiary of Bendix Corporation
Wiedemann Division, The Warner & Swasey Co., subsidiary of Bendix Corporation

For their critical comments and recommendations, the author is indebted to the following reviewers:

Thomas F. Ury
N/C Instructor
Pikes Peak Community College
Colorado Springs, Colorado

Gary A. Volk
Assistant Professor, N/C
Illinois Central College
East Peoria, Illinois

Clifford Oliver
Engineering and Technology Department
Chabot College
Hayward, California

Robert Swan
Chairman, Machine Shop Department
Asheville-Buncombe Technical College
Asheville, North Carolina

CHAPTER 1

Numerical Control — History and Evolution

OBJECTIVES   After studying this chapter, the student will be able to:
- Discuss the general history of numerical control.
- Understand the basic types of work performed on numerically controlled machines.
- Identify some basic types of numerically controlled machines.
- Discuss the general terms of accuracy, repeatability, and reliability as applied to N/C equipment.

GENERAL HISTORY OF NUMERICAL CONTROL

Perhaps the first real numerical control (N/C) machine was the 1725 knitting machine, which was controlled by sheets of punched cardboard. Then, around 1863, the player piano was introduced. The knitting machine and the player piano are considered the true forerunners of modern numerical control. Actually, numerical control as we know it began in 1947. John Parsons of the Parsons Corporation, based in Traverse City, Michigan, began experimenting with the idea of generating thru-axis curve data and using that data to control machine tool motions. Numerical control originated as Parsons discovered a way of coupling computer equipment with a jig borer.

In 1949, there was a demand for increased productivity for the U.S. Air Materiel Command as the parts for its airplanes and missiles became more complex. In addition, the designs were constantly being changed and revised. A contract was granted to the Parsons Corporation to search for a speedy production method. In 1951, the Massachusetts Institute of Technology (MIT) assumed the effort. In 1952 MIT successfully demonstrated a model of the N/C machine of today. The machine was a vertical spindle Cincinnati Hydrotel with a lab-constructed control unit. The machine successfully made parts with simultaneous thru-axis cutting tool movements. MIT actually coined the term *numerical control.*

In 1955 at the national machine tool show, commercial models of numerically controlled machines were displayed, ready for customer acceptance. Those shown were different contour-milling machines, which cost several hundred thousands of dollars. Some machines required trained and skilled mathematicians and computers to produce tapes. By 1957, numerically controlled machines were accepted by industry; several had been installed and were in use in some production applications.

Numerical control rapidly evolved in the control industry as well as the machine tool industry. At first, miniature electronic tubes were developed, and controls were big and bulky. Then came solid-state circuitry and eventually modular, or integrated, circuits. The size of the control units began to decrease. Also, controls became more reliable and less expensive. Experimental and developmental work continues on both machine tool and control unit capacity and design.

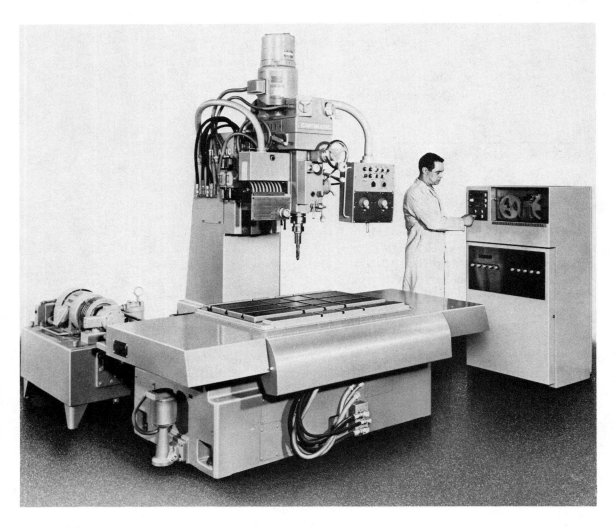

FIGURE 1-1
Early model of an N/C machine (Courtesy of Cincinnati Milacron Inc.)

TYPES OF MACHINES CONTROLLED BY N/C

At present, there are many types of N/C machines producing parts in manufacturing plants. They range from the earlier models of N/C machine tools, figure 1-1, to the advanced N/C profilers of today, figure 1-2. The sizes, capabilities, and options vary with each N/C machine. The one common factor is that they are N/C and can be programmed.

The most common types of N/C machines currently in use are of the chip-making variety such as N/C turning and machining centers, figures 1-3 and 1-4. Other types of numerical control machines include drafting machines, hole-punching machines, tube benders, inspection machines, riveting machines,

FIGURE 1-2
An advanced N/C profiler (Courtesy of Cincinnati Milacron Inc.)

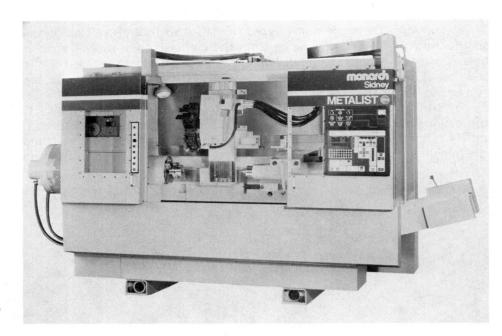

FIGURE 1-3
A modern N/C
turning center
(Courtesy of
Monarch Machine
Tool Company)

FIGURE 1-4
A typical N/C
machining center
(Courtesy of
Kearney & Trecker
Corporation)

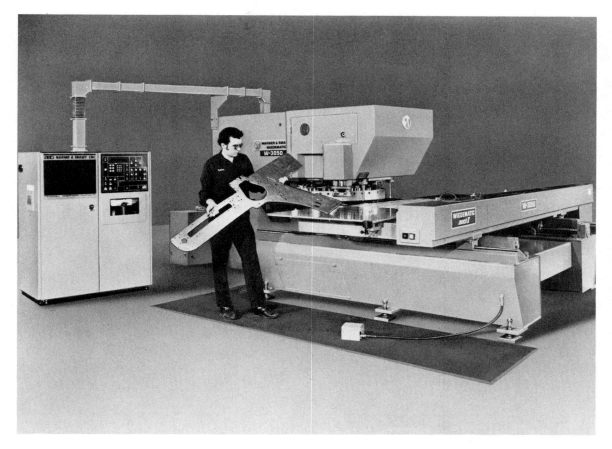

FIGURE 1-5
An N/C hole-punching machine (Courtesy of Wiedemann Division, The Warner & Swasey Co., subsidiary of Bendix Corporation)

welding machines, and flame cutters. Some examples are seen in figures 1-5 and 1-6.

Fortunately, a lot of the work needed to program the tremendous variety of modern machines and controls has been simplified, much to the credit of the manufacturers involved. The perforated or punched tape has become the major input medium. It is still considered an important N/C standard even though direct computer operation is the latest development, figure 1-7. Regardless of the type of machine N/C is controlling, the basic method of calling for machine action is the same for most types of N/C machines.

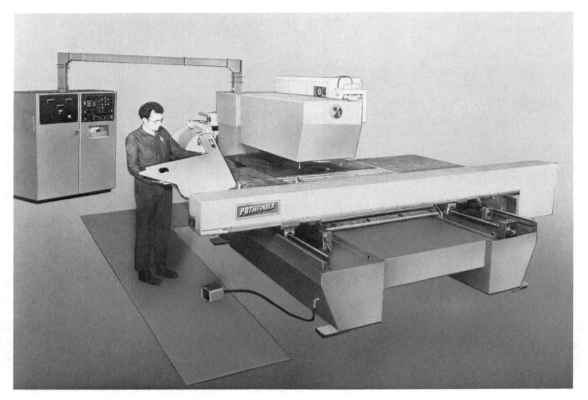

FIGURE 1-6
An N/C laser contour cutting system (Courtesy of Wiedemann Division, The Warner & Swasey Co., subsidiary of Bendix Corporation

ACCURACY, REPEATABILITY, AND RELIABILITY

Accuracy is defined as being free from error. The machinist may be able to produce a part that is free from error. Many machinists can work to a tolerance of .001 inch on an old, worn machine if they are very familiar with that machine. A tremendous amount of experience and time is needed to develop this degree of accuracy. Some N/C machines today can produce accurate workpieces with tolerances of ±.0002 inch or even ±.0001 inch. The operator, however, may not need to be familiar with the individual machine since the accuracy is built into the control and machine tool. These degrees of accuracy are not uncommon; some N/C machines can be purchased with systems capable of controlling much closer tolerances.

There are many factors which can affect a machine's accuracy. First, the foundation must be solid and conform to the manufacturer's specifications. Second, the lubrication schedule and all proper maintenance procedures must be strictly adhered to at all times. In addition, cutting loads and forces, temperature of the environment, material to be machined, types of

FIGURE 1-7
CNC mill (Courtesy of Bridgeport Machines)

cutting tools, and toolholders can greatly affect the accuracy of an N/C machine.

[Repeatability] is the comparison between the same dimensions of each piece machined. The repeatability of N/C is roughly about one-half of the actual positioning tolerance. The greater the accuracy and repeatability of the machine, the higher the cost. Repeatability is similar to accuracy in that the machine must receive proper care. The other factors affecting accuracy will also affect repeatability. Another element which should be pointed out is the care the operator must exercise in locating the parts in fixtures, vises, or whatever means are used to locate and clamp the workpiece. N/C operators should be aware that parts must be accurately located in the work-holding device, against positive stops. When the part is clamped, it is important to make sure it has not moved or been distorted out of position due to clamping forces. These simple checks ensure greater repeatability and quality of parts produced.

Reliability is another important goal of modern N/C manufacturers and users. The skilled hands of a good machinist are difficult to find; thus, the reliability that was previously built into the hands of the machinist must now be built into the N/C machine. Customers are demanding greater accuracy and reliability of the products. In order to meet and surpass this challenge, new types of slides, machine tables, bearings, and lead screws, are constantly being tried and tested.

The quality of a machine's parts and the tolerances with which they are manufactured and assembled are also important to machine reliability. Machine tool users, faced with rising costs, decreasing productivity, and increasing competition, must have a product they can depend on to produce accurate parts. The machine tools should have few maintenance or downtime problems. For this reason, reliability will continue to be of great importance to machine tool builders in future years.

REVIEW QUESTIONS

1. John Parsons is often considered the father of N/C. What educational institution helped in the refinements of N/C?
2. What conditions in the United States added to the development of N/C?
3. Why is it necessary to build the skill level into N/C machines?
4. Briefly explain the evolution of the N/C industry.
5. Name different types of N/C machines. Which types are the most common today?
6. To what degrees of accuracy are N/C machines able to produce?
7. Define repeatability. What is considered the general rule of thumb for determining the repeatability tolerances of an N/C machine?
8. What is the difference between accuracy and repeatability?

CHAPTER 2

What Is Numerical Control Programming?

OBJECTIVES After studying this chapter, the student will be able to:

- Discuss the importance of numerical control to manufacturing and productivity in the United States.
- Define numerical control.
- Explain the advantages and disadvantages of numerical control.
- Understand the fundamental steps of planning for the use of N/C.
- Describe general considerations and factors involved in N/C justification.

IMPORTANCE OF NUMERICAL CONTROL

In the early 1940s, manufacturing in this country was heading in one direction — mass production. The economy was demanding a large volume of goods at competitive prices. *Automation* was the key to satisfying the demands of the market. However, automatic machines were expensive, and large lot sizes were required for these machines to be justifiable. At that time, mass production was aimed at the median of consumer taste. This resulted in mass mediocrity and, consequently, consumer dissatisfaction.

In the 1960s, the market, coupled with numerical control and other new production technologies, turned around completely. People began demanding a variety of products from which to choose. This required more versatile production equipment and smaller production runs. Manufacturers were concerned because universal machines were now needed where highly specialized machines were previously used. These conditions placed a great deal of responsibility on numerical control technology and the promise it held for maximizing profits and increasing productivity.

So where does that leave us as to the importance of numerical control? Numerical control is more important now than ever. In manufacturing output, Japan and most of Europe have made monumental strides in recent years. As a result, they are gaining on U.S. productivity levels. In certain areas, foreign competitors have surpassed U.S. production levels.

The real importance of numerical control lies in the effects it has produced in this country. N/C machines are faster, more accurate, and more versatile where complex shapes are to be machined and where otherwise manual operations would be required.

Numerical control has risen in popularity through its ability to manufacture products of consistent quality more economically than alternative methods. Nevertheless, it is a popular misconception that numerical control is justifiable only for large-quantity production; just the opposite is true. An actual comparison of numerical control to standard machining methods indicates that the break-even point comes earlier with N/C production than with conventional production.

What exactly is numerical control, and what does it mean? Basically, it is control of machine tools by numbers. *Numerical control* is a system in which programmed numerical values are directly inserted, stored on some form of input medium, and automatically read and decoded to cause a corresponding movement in the machine which it is controlling.

NUMERICAL CONTROL — WHAT IT CONSISTS OF

Since numerical control is the control of machines by numbers, what numbers are used? How are they presented? These are only a few of the questions that are basic to an understanding of numerical control. These questions and many others will be answered in this text in the sequence necessary to lead to a complete understanding of numerical control.

Two important points should be made about N/C. First, the actual N/C machine tool can do nothing more than it was capable of doing before a control unit was joined to it. There are no new metal removing principles involved. N/C machines position and drive the cutting tools, but the same milling cutters, drills, taps, and other tools still perform the cutting operations. Cutting speeds, feeds, and tooling principles must still be adhered to. Given this knowledge, what is the real advantage of numerical control? Primarily, the idle time or time to move into position for new cuts is limited only by the machine's capacity to respond. Because the machine receives *commands* from the machine control unit (MCU), it responds without hesitation. The actual utilization rate or chip making rate is therefore much higher than on a manually operated machine. The second point is that numerical control machines can initiate nothing on their own. The machine accepts and responds to commands from the control unit. Even the control unit cannot think, judge, or reason. Without some input medium, e.g., punched tape or direct computer link, the machine and control unit will do nothing. The N/C machine will perform only when the N/C tape is prepared and loaded and cycle start is initiated.

ADVANTAGES AND DISADVANTAGES OF N/C

To begin a discussion of the advantages of numerical control, one must first realize that the average part spends only 5% of its manufacturing time on a machine in the shop. Only 1 1/2% of the time on a machine is spent actually cutting metal, as shown in figure 2-1. N/C actually works to reduce the *nonchip* making time. With N/C performing such manual functions as selecting spindle speeds, feed rates, coolant control, and automatic fixture indexing, it has become generally accepted that N/C machines are the most effective manufacturing developments for reducing the unit cost of production.

As foreign competitors continue to capture a larger share of the U.S. manufacturing market, management and labor become greatly affected by numerical control and its manufacturing systems. Some basic advantages of N/C have been mentioned already. However, a better understanding will come from the following explanations.

There are four basic phases that occur in most manufacturing. The first phase is *engineering* or the determination of the product's size, shape, tolerances, and material. The second phase is *process planning* which includes the decisions made concerning the selection of the manufacturing system from order of operations to inspection standards. The third phase is *economic planning*, which includes determining economic lot sizes, raw materials, and inventory analysis. The fourth is the *production* phase, including training of machine operators, machine setup, and actual machine operations.

In conventional manufacturing, the production phase was frequently the only phase considered when judging the advantages of new developments in metalworking. All phases must be considered when judging N/C machines. N/C tends to reduce the importance of the production phase in relation to the others. When skillfully employed, N/C provides cost savings throughout the entire manufacturing process.

Before numerical control capability existed engineers were severely limited in the designing of shapes using conventional machining. N/C makes it

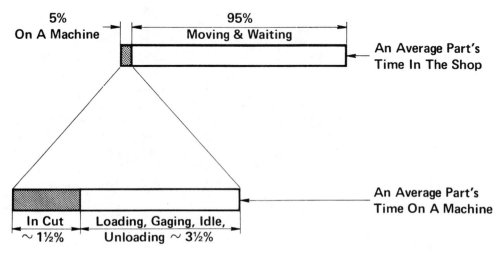

FIGURE 2-1
Breakdown of the time spent by an average part in the shop

possible to produce even the most complex shapes without extremely high costs. Another advantage of N/C is the ability to make changes or improvements with a minimum of delay and expense. With conventional machines, it is often economically undesirable to make changes after the tooling is prepared. In addition, costs associated with conventional machines increase as tolerances become tighter. This factor has caused engineering problems in trying to create parts with tolerances as loose as possible and still capable of functioning properly. With N/C, tolerances are somewhat independent of costs. The machine always produces parts to maximize accuracy without special treatment. (This is true if the operator locates parts properly, and so on.)

As discussed in Chapter 1, N/C machines provide good positional accuracy and repeatability. Complex jigs and fixtures are not required in all cases. For most operations, the simplest form of clamping devices is adequate. In addition to the reduction of complex fixtures, it is possible to reduce the use of expensive tooling. This factor greatly reduces the lead time required to get a new job into production.

Time study, in the conventional sense, is eliminated since the programmer now dictates how the part will be produced and how long it will take. After the program is established, there can be no variations from part to part and no deviation from the programmed time.

A high degree of quality is inherent in the N/C process because of its accuracy, repeatability, and freedom from operator-introduced variations. In-process quality inspection is seldom required after an inspection of the first part produced from a new tape, as a check on the programming function. A coordinate measuring machine, figure 2-2, is used to check positional accuracy.

One of the basic functions of economic planning, as previously suggested, is the determination of economic lot size. With conventional machining methods, setup costs are high and cannot be calculated with any degree of accuracy. Therefore, it is necessary to make a large number of parts for each setup if the unit part cost is to be minimized. With numerically controlled machines, the high process predictability ensures accurate cost determinations, and the simplified, low-cost setups enable parts to be run in small quantities economically.

In addition, since the programmer selects the methods and the sequence of operations, as well as operating feeds and speeds, cutting conditions are under the complete control of manufacturing supervision. With N/C, actual physical manipulation of the machine by the operator is greatly reduced since feeds and speeds are, in most cases, automatically selected.

Some other advantages of numerical control are as follows:

- reduced scrap. Errors due to operator fatigue, interruptions, and other factors are less likely to occur on N/C machines.

- improved production planning. N/C machines can often perform, at one setting, work that would normally require several conventional machines.

FIGURE 2-2
A coordinate measuring machine (Courtesy of Portage Machine Company)

- reduced space requirements. Since fewer jigs and fixtures are used, the actual storage requirements of these expensive tools are reduced.
- simplified inspection. Once the first piece has passed inspection, minimal inspection is required on subsequent parts.
- lower tooling costs. There is less need for complex jigs and fixtures.
- reduced lead time. This is a result of lower tooling costs.
- complex machining operations are more easily accomplished. This is due to advanced machine control and programming capabilities.

There are, however, some factors relating to numerical control which some individuals might call disadvantages or inhibitors to using numerically controlled equipment. Some of these disadvantages are worth mentioning, but a detailed analysis will, in most cases, reveal that the advantages of numerical control outweigh the disadvantages.

First, tools on N/C machines will not cut metal any faster than tools on conventional machines. N/C machines merely position and drive the cutting tools. Optimized feeds and speeds can be run on either conventional or N/C machines.

N/C does not eliminate the need for expensive tools. Some jobs require special and expensive fixtures and cutting tools. The most significant factor is the greater initial cost of the N/C machine, compared to that of a conventional machine. Machines and tooling are costly today, and their purchase requires extensive justification.

Another factor which must be considered is that, contrary to popular belief, N/C will not totally eliminate errors. Operators can still push the wrong buttons, make incorrect alignments, and fail to locate parts properly in a fixture. Some of these types of errors can be minimized by careful and effective training. However, some errors will always be likely to occur; they will never be totally eliminated.

The proper selection and training of programmers and maintenance personnel is required. The support personnel are essential to the success of any N/C installation, and must be given careful and adequate consideration.

These items may not be considered disadvantages of N/C as much as inhibitors to purchase. Undoubtedly, many smaller companies have decided not to purchase N/C equipment after weighing all costs and requirements involved. Like any advanced technological equipment, N/C should be used only where it will produce the work better, faster, and more accurately than conventional methods. Many shops, after reluctantly purchasing their first piece of N/C equipment, have found the actual savings and advantages to be much greater than originally planned.

PLANNING FOR THE USE OF N/C

What are the actual steps a potential N/C user takes prior to purchasing N/C equipment? There are ten factors which should be considered before using numerically controlled equipment. Depending upon the requirements of the facility, equipment to be purchased, and type of work to be produced, additional consideration may be necessary.

BASIC KNOWLEDGE OF HOW N/C WORKS

For any prospective user, there is no substitute for firsthand information concerning N/C machine and control unit operations and capabilities. One of the best ways in which potential N/C users can obtain this information is by attending N/C training programs and demonstrations offered by machine tool manufacturers and suppliers.

CAPITAL INVESTMENT REQUIREMENTS

As previously discussed, N/C machines require a substantial initial investment in comparison to standard machines. In numerical control, as applied to basic machine tools, the basic machine must be designed with a more rigid frame and heavier lead screws, bearings, and other actuating mechanisms. These allow the machine tool to achieve the acceleration speeds required for efficient positioning times. Added to the initial cost of the machine are tape preparation and data processing expenditures, as well as other costs necessary to maintain an N/C environment. The net capital expenditure may be in the range of $100,000 to $250,000 for an N/C lathe or machining center to more than $2,000,000 to $3,000,000 for highly sophisticated five-axis profilers. A careful survey will reveal that despite these high initial costs, a properly operated N/C installation will pay for itself in a remarkably short period of time.

PERSONNEL AND TRAINING

One of the most important points to keep in mind when considering the purchase of N/C machinery is that, in order to get the maximum return from the capital invested, good cooperation and communication must be maintained among shop, programming, and engineering personnel. N/C opens up great opportunities for the engineer and designer to use design techniques that were impossible when only conventional manufacturing methods were available.

When selecting personnel for operator, programmer, and maintenance positions, each company should survey its current employees for essential skills that these high-technology jobs require. It may be necessary for a company to hire outside its establishment rather than train existing personnel in order to obtain the required skills. Selecting qualified and skilled personnel is critical to the success of an N/C installation. This selection process must be approached cautiously and objectively; considerable thought, discussion, and research must be conducted prior to the installation of equipment.

PROGRAMMING AND TOOLING

One hurdle that management must clear in any consideration of N/C equipment is how to handle programming and preparation of tapes for the control unit. Later chapters will discuss the data processing equipment necessary to support the programming function and whether manual or computer methods of programming should be used.

Tooling for N/C is closely related to programming since it involves the choice, size, and shape of cutters to match the plan visualized by the programmer. It is customary for the program to include tool specifications so that the machine operator can get the complete set of tools for a given part without initially having to work out a required list. The type of N/C system under consideration will determine whether the programmer will use standard tools from a tool store or whether an elaborate tool library must be established for preset tools.

MAINTENANCE AND REPAIR

Another concern to management is maintenance. Some companies have been discouraged from installing N/C machines because they felt they did not have adequate facilities or personnel to service the electronically controlled equipment. With the advent of integrated circuits, microprocessors, and computer numerical control *(CNC)*, there has been a general increase in the reliability of controls. Still, the lack of trained or capable personnel is a concern for potential N/C users. However, maintenance classes are offered by control manufacturers to educate new maintenance personnel and update current personnel. It should be noted that maintenance, as well as programming and operation of the equipment, must be restricted to trained and authorized personnel only for purposes of safety and equipment protection. In order to facilitate maintenance, the electronic controls are often assembled in modular units. By keeping spares of the major units, time loss for correcting malfunctions within controls can be reduced to a minimum.

COST ANALYSIS

Generally, when discussing the cost analysis of numerical control, the programming costs and tape preparation time is being compared to the design and upkeep of the jigs, fixtures, and setups required for conventional machining operations. The elimination of elaborate jigs and fixtures probably constitutes the largest area of savings in numerical control.

Another item of importance in cost analysis is the shorter lead time of parts machined on N/C. Again, a typical part will spend about 5% of its manufacturing life being machined and the other 95% sitting in flats, being inspected, and waiting. With N/C, parts are routed to fewer machines, thus cutting down the total manufacturing time.

QUALITY CONTROL

One benefit of numerical control is the repeatability of parts produced and the reduced inspection time. N/C has made it possible to produce part after part with consistent accuracy. The adverse effects of operator skill, fatigue, and human reliability have been reduced to a minimum. More complex parts can now be produced with much lower rejection rates than conventional methods.

REDUCTION OF INVENTORY

With the advent of N/C, shorter lead times are needed, thereby reducing the total amount of inventory required. Finished inventory can be held to an absolute minimum because of the care and repeatability with which a part can be put in process and the resulting speed with which inventory can be replaced. Raw material inventories can also be cut since it is no longer necessary to schedule long runs.

ENVIRONMENTAL REQUIREMENTS

One of the most important factors to be considered in studying a plant layout for an N/C installation is accessibility. The flow of work to and from the machines is important due to the tremendous appetite of N/C machines. All plant layouts should be based on the maximum use factor of these units. Other environmental considerations should be given to cleanliness and possible extensive chilling or heating conditions.

SERVICE RESPONSIBILITY

As greater emphasis is placed on numerical control systems for manufacturing, it is essential that these systems be kept in operation as much as possible. It is often difficult to determine whether a problem originated with the control or the machine tool. This sometimes results in a situation where the user does not know whether to contact the machine tool service representative or the control service representative. Consequently, much time can be wasted trying to determine the source of the problem. Many companies will purchase both machine and control from the same manufacturer, thereby eliminating the question of who will make the repair.

N/C JUSTIFICATION

N/C justification has already been discussed in relation to the advantages of numerical control and planning for its use. However, some additional concepts should be examined.

At present, there are several approaches to equipment justification. One method is the cost-savings approach. It has the distinct advantage of being easy to calculate. It is basically conservative, based on equipment replacement with some degree of improved productivity and performance. The disadvantage to this approach is that there is no disciplined effort to review the entire operation for improvement.

Another widely used justification technique is aggressive justification. This method makes an advantage of the cost-savings approach disadvantage. Aggressive justification questions whether the present methods are the best, and it may involve substantial changes in manufacturing methodology. There are tremendous opportunities for processing improvements with this method, but the translation into dollars and cents is complex.

Whenever an N/C justification is needed, an analysis of the parts to be programmed is an ideal place to start. Once it is determined that a realistic N/C work load exists, the second step is to determine the return on investment (ROI). In calculating the ROI, it is necessary to study the different components of the business that are affected by the use of this new equipment, not just the machine itself. One of the questions that should be answered is, "Will N/C help produce a better and consistent quality part in an economical manner?" The calculation is computed by dividing the average savings per year by the initial investment cost. The result is a rate of return.

$$\frac{\text{Avg. savings/yr.}}{\text{Investment}} = \text{Rate of return}$$

Perhaps some productivity comparisons will also have to be made by estimated production lot size, figures 2-3 and 2-4.

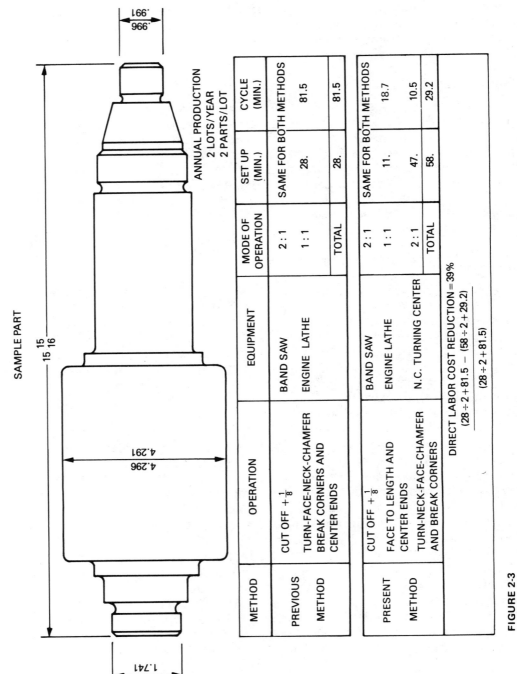

FIGURE 2-3
Productivity comparison — small lot sizes

SAMPLE PART

ANNUAL PRODUCTION
2 LOTS/YEAR
2 PARTS/LOT

METHOD	OPERATION	EQUIPMENT	MODE OF OPERATION	SET UP (MIN.)	CYCLE (MIN.)
PREVIOUS METHOD	CUT OFF $+\frac{1}{8}$	BAND SAW	2 : 1	SAME FOR BOTH METHODS	
	TURN-FACE-NECK-CHAMFER BREAK CORNERS AND CENTER ENDS	ENGINE LATHE	1 : 1	28.	81.5
			TOTAL	28.	81.5
PRESENT METHOD	CUT OFF $+\frac{1}{8}$	BAND SAW	2 : 1	SAME FOR BOTH METHODS	
	FACE TO LENGTH AND CENTER ENDS	ENGINE LATHE	1 : 1	11.	18.7
	TURN-NECK-FACE-CHAMFER AND BREAK CORNERS	N.C. TURNING CENTER	2 : 1	47.	10.5
			TOTAL	58.	29.2

DIRECT LABOR COST REDUCTION = 39%

$$\frac{(28 \div 2 + 81.5) - (58 \div 2 + 29.2)}{(28 \div 2 + 81.5)}$$

SAMPLE PART

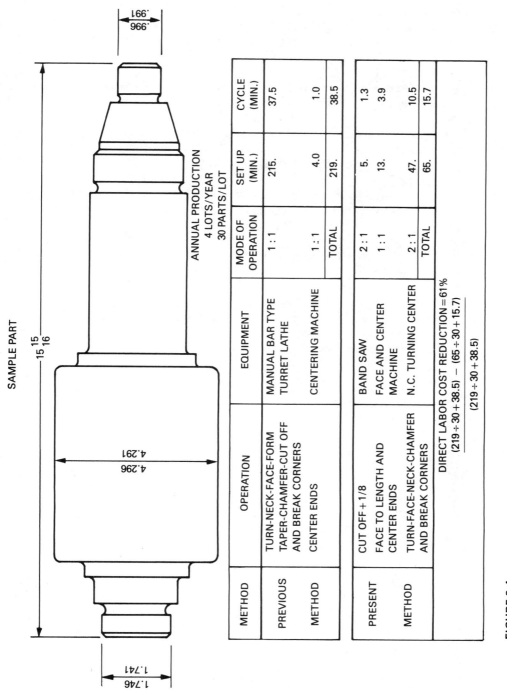

ANNUAL PRODUCTION
4 LOTS/YEAR
30 PARTS/LOT

METHOD		OPERATION	EQUIPMENT	MODE OF OPERATION	SET UP (MIN.)	CYCLE (MIN.)
PREVIOUS METHOD		TURN-NECK-FACE-FORM TAPER-CHAMFER-CUT OFF AND BREAK CORNERS	MANUAL BAR TYPE TURRET LATHE	1 : 1	215.	37.5
		CENTER ENDS	CENTERING MACHINE	1 : 1	4.0	1.0
				TOTAL	219.	38.5
PRESENT METHOD		CUT OFF + 1/8	BAND SAW	2 : 1	5.	1.3
		FACE TO LENGTH AND CENTER ENDS	FACE AND CENTER MACHINE	1 : 1	13.	3.9
		TURN-FACE-NECK-CHAMFER AND BREAK CORNERS	N.C. TURNING CENTER	2 : 1	47.	10.5
				TOTAL	65.	15.7

$$\text{DIRECT LABOR COST REDUCTION} = 61\% \quad \frac{(219 \div 30 + 38.5) - (65 \div 30 + 15.7)}{(219 \div 30 + 38.5)}$$

FIGURE 2-4
Productivity comparison — production lot sizes

Regardless of the method of N/C justification that is used, much more detailed work and analysis is needed than is mentioned in this chapter. These are merely two of the many different types of N/C justification techniques.

REVIEW QUESTIONS

1. What were some of the important factors leading to the popularity of numerical control?
2. In your own words, briefly define numerical control.
3. The average part spends only 5% of its manufacturing time on a machine. What comprises the other 95% of an average part's time on the shop floor?
4. What are the four basic phases that occur in most manufacturing? Briefly explain their importance and relationship to one another.
5. List and briefly describe the major advantages of numerical control. What are the disadvantages or inhibitors to numerical control?
6. What is the major advantage of numerical control over conventional machining?
7. List and briefly describe the ten basic considerations in planning for the use of N/C.
8. Name two techniques used for successful N/C justification, and briefly explain their functions. Which approach is most likely to yield the greatest overall savings? Why?

CHAPTER 3

How Numerical Controls Operate

OBJECTIVES After studying this chapter, the student will be able to:

- Explain how a workpiece is processed for numerical control application.
- Understand the function of tape readers.
- Describe axis relationships and tape readout characteristics.
- Explain how numerical controls collect and store information.
- Compare the different types of numerical control systems.
- Discuss the common types of N/C feedback systems and their primary purpose.

WHAT A MACHINIST NEEDS TO KNOW

Since the number of skilled machinists in this country is on the decline, these skills must be compensated for in another manner. The machine and control unit, through the direction of the programmer, now perform many of the functions of the skilled machinist.

The entire process of N/C programming is being able to visualize the actual cutting motions and table movements that are taking place on the machine. These movements are then translated into a coded *format*. Consequently, many interactive decisions and judgments are made by programmers as well as machinists.

Let us study some of the judgments and functions of a machinist. A machinist will begin a workpiece with a thorough study of a blueprint, sketch, or sample workpiece. In addition, a machinist will check the record or process routing sheet, if available, to determine the specific machining to be performed for that particular operation. If specific tolerance requirements are needed, such as an allowance for grind stock, the process sheet should contain this information for the machinist's review prior to machining. Another important step in machining a workpiece is planning the sequence of machining operations. The concept of "What do I do first; what do I do next?", along with determining how the individual setup is to be made, is of prime importance. These decisions are closely related to the experience level of the machinist. A machinist also calculates the speeds and feeds, and selects cutting tools, materials, and machine tools where appropriate.

There may be more detailed items that an individual machinist must be concerned with, but essentially the machinist has visualized a *program*. This program tells what and how a particular workpiece is to be machined. The machinist then guides the cutting tools in their relationship to the workpiece by means of the operating dials and levers. After the machining operations are complete and the material is removed from the workpiece, an accurate part is produced. In essence, a skilled machinist has programmed, accumulated, stored, and transmitted detailed information in order to produce a particular workpiece.

WHAT A PROGRAMMER NEEDS TO KNOW

A *programmer* must use all the skills of a machinist in processing a part. In addition, a programmer must study the blueprint and process sheet, know how to set up the workpiece, and be familiar with the sequence of machining operations. A programmer's selection of cutting tools, feeds and speeds, and basic programming methodology are critical to efficient use of the machine used to process the part. In addition to visualizing the entire process, the programmer must also be familiar with the particular machine being programmed and its general operating characteristics and requirements. The programmer must also understand the specific tape format that the particular machine control unit will accept. An understanding of computer programming languages may be necessary to obtain the desired output results. Essentially, the programmer has accumulated, stored, and transmitted data, resulting in the completed N/C program. A flow diagram of the steps in processing an N/C program is illustrated in figure 3-1.

WHAT MAKES A GOOD N/C PROGRAMMER?

A clear description of an N/C programmer may be necessary to answer this question. Some companies may require a programmer to write an N/C *code* according to a prepared production plan, using previously selected tooling and fixtures. Other firms may require the programmer to plan the sequence of machining operations, selecting the appropriate tools and fixtures, as well as preparing the N/C coding. Actually, an N/C programmer can vary from being a coder to a comprehensive manufacturing engineer.

One of the most important qualities of a successful N/C programmer is skill in print reading. The programmer must be able to visualize, in three dimensions, parts depicted on a two-dimensional blueprint. Another desirable quality, but one that is becoming increasingly more difficult to obtain, is time on the shop floor solving the multitude of manufacturing problems. This type of individual represents the optimum mix.

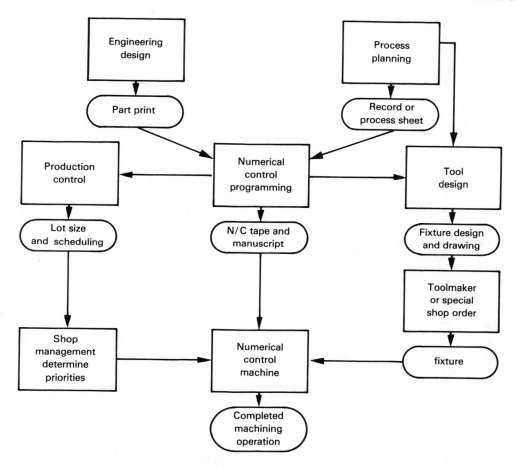

FIGURE 3-1
A flow diagram of the steps in processing an N/C program

Many talented individuals are currently being trained by colleges and universities. These individuals have the related academic skills and, with a little patience and effort on the part of manufacturers, they can obtain the necessary manufacturing experience needed to become competent and successful programmers.

HOW N/C COLLECTS INFORMATION

Information is passed from the N/C *tape* to the machine control unit by means of the tape *reader.* A photoelectric tape reader and the actual loading of an N/C tape into a tape reader is shown in figures 3-2 and 3-3. Tape

FIGURE 3-2
A typical numerical control cabinet with a tape reader and tape reels (Courtesy of Cincinnati Milacron Inc.)

readers generally are classified as mechanical or photoelectric (light). They may read a single row of information at a time, or they may read a complete *block* of instruction. Reader speeds will vary considerably; mechanical readers are capable of reading approximately sixty characters per second, and photoelectric readers can read approximately 300 to 500 characters per second.

Mechanical readers, sometimes called electro-mechanical readers, are used mainly on point-to-point N/C systems, although limited contouring work may be done with mechanical readers. The advantages of mechanical readers are that they are extremely reliable, easy to maintain, and relatively

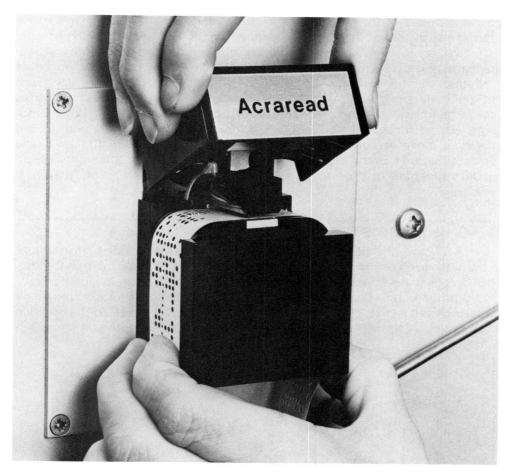

FIGURE 3-3
Loading a tape reader (Courtesy of Cincinnati Milacron Inc.)

inexpensive. Their slower reading speeds, however, present a prime disadvantage in some contouring situations. This occurs when hundreds of *characters* are needed to complete a large radius or other configuration. Mechanical readers operate by mechanically sensing, using pins, the codes punched into the tape. When the pins enter a hole present in the tape, contact is made and an impulse is created that is transmitted to the control unit. The sprocket on the tape reader actually feeds the tape over these pins and, at the same time, guides the tape through the reader.

Photoelectric readers are the most commonly used because of their speed. They shorten the time between reading and performance. Photoelectric readers operate by light beams which pass through the holes in the tape and impinge on a photocell. The light beams are then converted to electrical impulses and passed on to the controller. Light readers have no drive sprockets to drive the tape; therefore, direct drive or rotating/braking capstans are used. The actual braking and accelerating are almost instantaneous, and the reading speed of 300 to 500 characters per second enhances the smooth and continuous motion of the machine. The only drawback to photoelectric readers is that they are very sensitive to dirt, unaligned sprocket holes, and imperfections in tape and hole punching. These problems can, at times, cause erroneous tape readings. Consequently, a displaced machine move or command will occur, or the control will display a tape error situation. Every effort should be made to ensure that only clean, quality tapes are used on machine control units equipped with photoelectric readers.

MACHINE REGISTERS AND BUFFER STORAGE

After the tape has been read by the tape reader, the coded information, now in the form of signals, is passed on to the control. *Registers* within the control accept the information, which consists of proper coordinates, preparatory functions, and miscellaneous functions. This information is transmitted to the respective register sections where actuation signals are relayed to the machine tool drives. A fundamental sketch of this process is illustrated in figure 3-4.

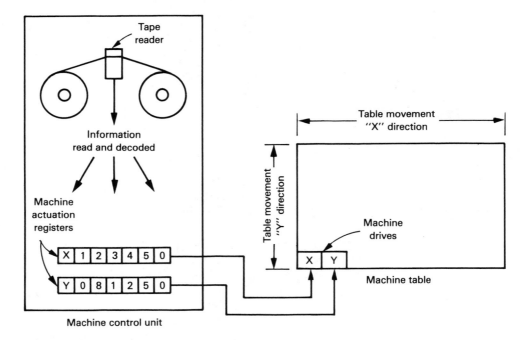

FIGURE 3-4
Fundamental sketch of tape input information being read, decoded, and passed to machine actuation registers, resulting in corresponding table movements

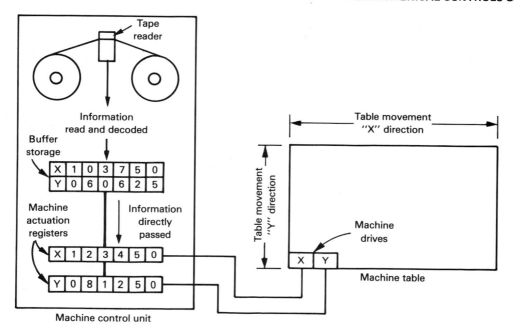

FIGURE 3-5
Fundamental sketch of tape input information being read, decoded, and stored in buffer storage until machine actuation registers have completed previous move and commands. Information is then passed from buffer storage to actuation registers.

Most modern N/C and CNC controls are equipped with *buffer storage.* As shown in figure 3-5, this feature allows the control to accept information into a buffer register while an operation is being performed from the active machine registers. When that operation is completed, the information is transferred from buffer storage to the machine actuation registers. This transfer of information is instantaneous, thereby reducing the time between tape reading and machine performance. Buffer storage reduces the amount of *dwell time* between machine operations because the next block of tape is read and stored while the machine is executing the previous block. Part finish is also better because the cutter does not come to a halt to process the next block of information in the middle of curves, angles, or other part configurations.

AXIS RELATIONSHIPS — READOUT

Once the N/C tape has been read, decoded, and the machine actuation registers loaded, the machine responds with its appropriate coordinate axis movements and other commands. The primary axis movements to be studied in terms of their relationship to each other are the X and Y axes. *Axis* refers to any direction of motion which is totally controlled by specific tape commands. Machines with only X and Y positioning capability are known as

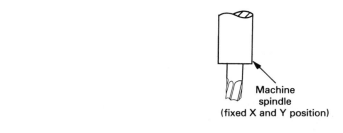

Machine
spindle
(fixed X and Y position)

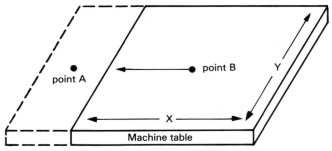

FIGURE 3-6
A machine table movement to the left in the X
direction is needed to move from point A to
point B.

two-axis machines. On machines of this type, distances in the Z direction
are controlled by the operator or by preset stops similar to that of a conventional drill press. On some two-axis N/C machines, the Z depth is controlled
by a system of *cams* or operator-set micrometer stops which can then be
selected by tape command. An example of table movements in the X-Y plane
is shown in figure 3-6. The three basic motions, as designated by the Electronic Industries Association (EIA), are X, Y, and Z. The principle X
motion is parallel to the longest dimension of the primary machine table. The
primary Y movement is normally parallel to the shortest dimension of the
primary machine table. The Z motion is the movement that advances or
retracts the spindle. The Z axis movement becomes complicated by the fact
that N/C machines are made with vertical and horizontal spindles. To help
understand the Z axis, it can be said that a line through the center of the
machine spindle is actually the Z axis. It is only when the actual depth of
cut (Z axis) is controlled by the tape that the machine is considered a true
three-axis N/C machine. That is, the machine is capable of *simultaneous*
motion in X, Y, and Z. As will be shown later, there are N/C machines
which will call for a set of preset depths and Z motion. These are not
considered true three-axis machines even though the machine does accept
tape-actuated Z motions. Explanation of the machine axes in relation to
coordinate positions will be detailed in Chapter 4. Additional machine/axis
relationships can be seen in figures 3-7, 3-8, 3-9, and 3-10.

As the N/C machine positions itself, corresponding to the programmed
positions in the tape, the positions obtained are displayed by tape command
readouts. The readouts are generally for *sequence number* and X, Y, and
sometimes Z axes, although additional information can be displayed on more

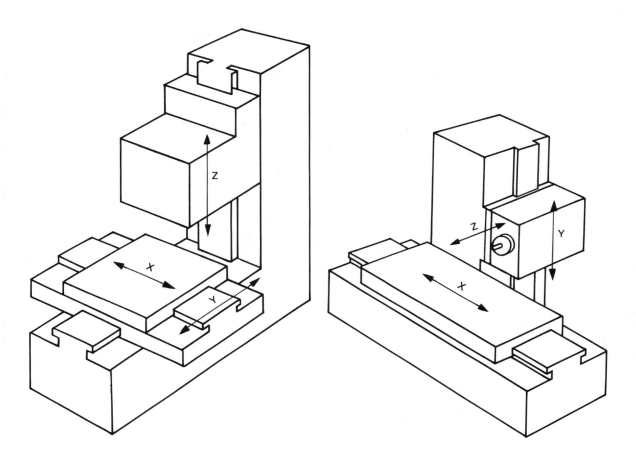

FIGURE 3-7
A vertical N/C machine

FIGURE 3-8
A horizontal N/C machine

sophisticated controls. These readouts consist of lighted numbers mounted in the machine control unit and/or on the operator's console or *display* screen.

The individual sequence numbers identify the block or line of information being read. The X- and Y-positioned readouts are constantly changing as the program advances, and are mainly provided so that the operator can identify specific positions and lines or blocks of information relative to the program manuscript.

TYPES OF FEEDBACK SYSTEMS

Once the command signals have been sent from the machine control unit to the machine, the slide motion and spindle movement occurs. How

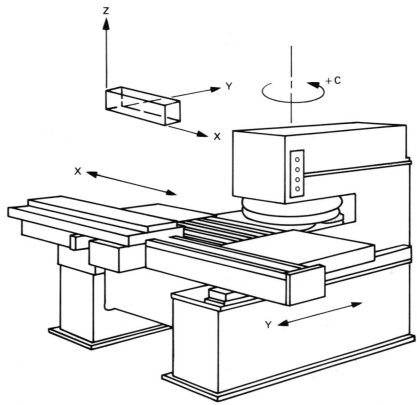

FIGURE 3-9
An N/C turret punch press

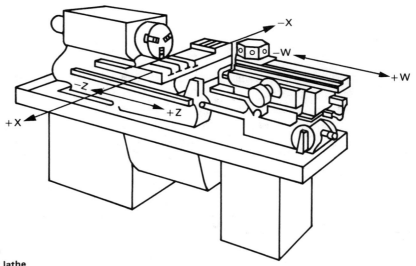

FIGURE 3-10
An N/C turret lathe

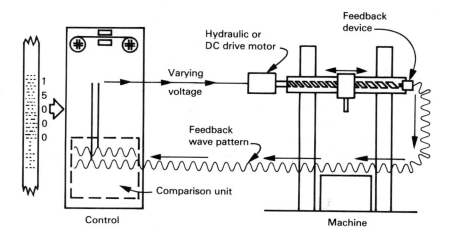

FIGURE 3-11
Closed loop

then does the N/C control know that the machine is properly positioned? Unless the control and machine form a closed-loop system, the control really has no way of knowing if the machine is properly positioned. Figure 3-11 illustrates a basic closed-loop system. *Closed-loop systems* are similar in operation to driving an automobile. When driving an automobile, an individual will check the speedometer to determine speed. The actual speed is compared to the desired speed which is designated by the speed limit signs. The driver's brain detects the difference between the posted speed and the actual speed, and the brain then instructs the foot to adjust the condition until the correct speed has been achieved.

Open-loop systems provide no check or measurement to indicate that a specific position has actually been achieved. No feedback information is passed from the machine back to the control. The system components may be affected by time, temperature, humidity, or lubrication, and the actual output may vary from the desired output. The main difference between open- and closed-loop systems is that with closed-loop systems, the actual output is measured, and a signal corresponding to this output is fed back to the input station where it is compared to the input registers. Such a system automatically attempts to correct any discrepancy between desired and actual output.

Feedback systems may be either *digital* or *analog*. Digital systems generate pulses which are fed back to the control and count down linear or rotary motion in minimum movements on machine lead screws. Analog systems sense and monitor variations in levels of voltage. Moving tables on a machine may overshoot in both directions and then search for the exact position to stop.

TYPES OF NUMERICAL CONTROL SYSTEMS

Numerical control systems are generally classified into two basic types: positioning or point-to-point; and continuous-path or contouring.

Positioning, or *point-to-point programming,* as illustrated in figure 3-12, is best described as moving or directing a tool to a specific location on a workpiece to perform operations such as drilling, tapping, boring, reaming, and punching. The process of positioning from one coordinate (X, Y) position to another, performing these basic operations, continues until all work has been performed for programmed locations. The important aspect of positioning systems is that, on a true positioning system, the cutting tool is never in constant contact with the workpiece. The spindle or the table may move to locate the desired position directly under the spindle. When the X and Y position is satisfied, the spindle will then advance the cutting tool into the workpiece. There are some positioning systems that do possess limited contouring capabilities such as straight-line milling along either the X or Y axis. In addition, 45-degree-angle milling is possible on some positioning systems with limited contouring capabilities.

Contouring, or *continuous-path, systems* maintain a constant cutter-workpiece relationship. The cutting tool remains in constant contact with the workpiece as the corresponding coordinate movements are attained. This process is illustrated in figure 3-13. The most common of the continuous-path operations are milling and lathe operations which profile workpieces to exact specifications. The actual contouring control system must have speed control independent of the X and Y driving motors. This enables the rate of travel to be regulated on at least two axes at the same time.

The distinction between positioning and contouring control systems has significantly decreased; the majority of control units possess both positioning

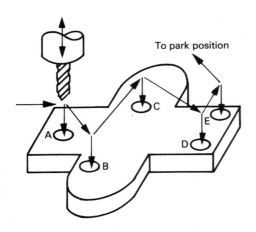

FIGURE 3-12
Positioning

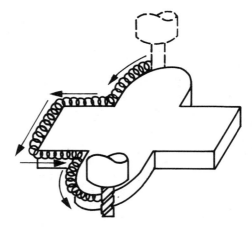

FIGURE 3-13
Contouring

and contouring capabilities. Control units have advanced this far because of the progress made in sophisticated electronics, and the need due to workpiece complexity, tighter tolerances, and improved part finishes. In addition, the requirements needed to program and operate these advanced controls have increased substantially, resulting in greater usage of available computer languages.

REVIEW QUESTIONS

1. Compare a machinist's and programmer's processes of producing a workpiece.
2. What are the two major types of tape readers available? What are the advantages and disadvantages of each?
3. Describe in detail the kinds of applications in which slow reading speeds are a drawback?
4. What type of tape reader is best suited for contouring applications?
5. Describe how buffer storage operates. What is its main advantage?
6. How is the Z axis on an N/C machine determined?
7. Discuss the basic process which occurs within the control unit between the tape being read and actual machine movement.
8. What are the primary differences between point-to-point and contouring N/C systems?
9. What is the difference between a closed-loop system and an open-loop feedback system?
10. In your own words, describe the function of a closed-loop system.
11. What is the primary advantage of a closed-loop system?
12. Explain why the distinction between positioning and contouring N/C systems has become less significant.

CHAPTER 4

Rectangular Coordinates — Absolute and Incremental

OBJECTIVES After studying this chapter, the student will be able to:

- Understand the Cartesian coordinate system in conjunction with numerical control coordinate systems and machine axes.
- Discuss absolute and incremental dimensioning and control systems.
- Understand the difference between fixed zero and full floating zero control systems.
- Explain the advantages and capabilities of floating zero systems.

CARTESIAN COORDINATE SYSTEM

Numerical control is based on the principle of rectangular coordinates discovered by the French philosopher and mathematician, René Descartes. This mathematical development of over three hundred years ago is better known as the Cartesian coordinate system. Through the use of rectangular coordinates, any specific point in space can be described in mathematical terms along three perpendicular axes. The idea relates perfectly with machine tools since their construction is generally based upon two or three perpendicular axes of motion.

The system of *Cartesian coordinates* is illustrated in figure 4-1. As in algebra, the *X axis* is horizontal (left and right), and the *Y axis* is vertical (up and down). The *Z axis* is then applied by holding a pencil perpendicular to the paper with its point at the location where the X and Y lines cross each other. The point where the X and Y axes cross is called the origin or zero point. Four quadrants are formed when the X and Y axes cross. Each *quadrant* is numbered in counterclockwise rotation (figure 4-1).

The plus (+) and minus (−) signs indicate a direction from the zero point along the X and Y areas. As seen from figure 4-1, all positions plotted in the first quadrant have positive X and positive Y values (+X,+Y). All positions in the second quadrant have negative X and positive Y values (−X, +Y). Points plotted in the third quadrant have values of negative X and negative Y (−X,−Y). In the fourth quadrant, all positions have values of positive X and negative Y (+X,−Y).

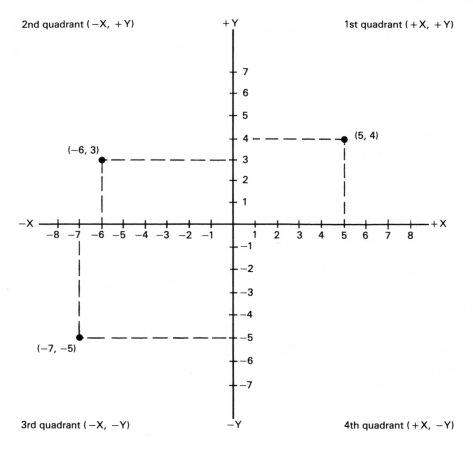

FIGURE 4-1
Cartesian coordinate system

At first glance, it appears it would be easier if all work could be done in the first quadrant since all values are positive and no signs would be needed. However, any of the four quadrants may be used on different machines. In some applications, the use of minus signs is a distinct advantage. Therefore, programmers must be thoroughly familiar with the use of both plus and minus signs in all four quadrants.

TWO-AXIS TAPE CONTROL

Two-axis tape control normally consists of the X and Y axes. In figure 4-1, it can be seen that the point labeled (5,4) actually means X = 5.0000 and Y = 4.0000 from the zero point. The point (−7, −5) means X = −7.0000 and Y = −5.0000 in relation to the zero point. It is generally agreed that in this kind of notation, the X dimension is written first, followed by the Y dimension (X,Y).

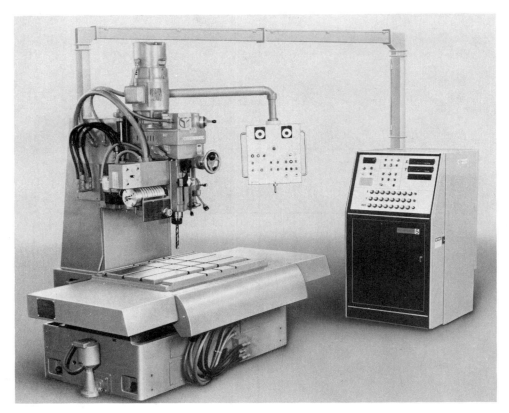

FIGURE 4-2
A typical two-axis N/C machine

A two-axis N/C machine is shown in figure 4-2. In most cases, the machine table moves left and right to position the cutter. A longitudinal movement of the machine table on this machine is an X axis move. A cross or saddle movement is a Y axis move. In order to obtain identical X-Y movements, on some two-axis machines the table remains stationary, and the spindle moves to satisfy X and Y locations. The movement of the cutter remains the same regardless of whether the spindle or table positions the workpiece. This consideration is taken care of by the machine and control manufacturers. The programmer must specify only the dimensions and the plus or minus signs in relation to the zero point. How the coordinate system relates to the machine table will be explained in detail later.

In numerical control programming, all plus signs may be written beside all positive locations. However, to make the work easier, a plus sign does not need to be written if the value is positive unless it is required by that particular N/C machine. In all cases, minus signs *must* be written to distinguish between negative and positive values when the sign has been omitted.

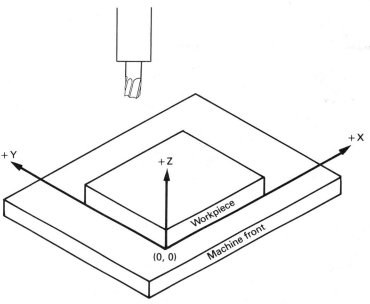

FIGURE 4-3
Vertical Z axis drawn in relation to the X and Y axes, workpiece, and machine table

---

Z AXIS CONTROL

Any type of advanced work in numerical control will involve the Z axis. As was explained earlier, a line through the center of the machine spindle is the Z axis. The principle applies regardless of the machine type. Z axis motions can be operator-controlled with preset stops or programmer-controlled through the N/C tape. Figure 4-3 illustrates a vertical Z axis with the X and Y axes drawn in relation to the workpiece and machine table. A positive Z movement is described as moving the tool *away* from the work. A negative Z movement is described as a plunge cut or moving the tool *into* the workpiece. Figure 4-4 illustrates a horizontal Z axis, with the X and Y axes drawn in relation to the workpiece and machine table. Positive Z movement moves the tool *away* from the work, and negative Z movement moves the tool *into* the work. Notice that as the workpiece is positioned in relation to the zero point, all positions on the XY plane surface will have positive values and will be situated in the first quadrant.

FOUR- AND FIVE-AXIS TAPE CONTROL

Some of the most advanced N/C machines are equipped with four- and five-axis contouring capabilities. These capabilities are generally secondary and tertiary axes which parallel the X, Y, and Z axes, plus some additional

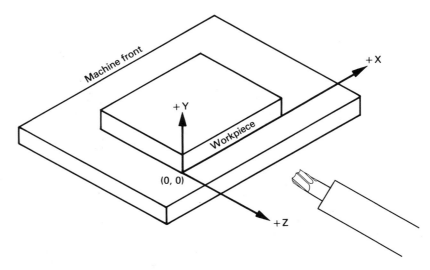

FIGURE 4-4
Horizontal Z axis drawn in relation to the X and Y axes, workpiece, and machine table

rotational movements within the coordinate locations. The rotational movements can usually sweep through all axes in complex cutting tool movements and applications. These other axes are given letters. Their programming becomes quite complex even with advanced computer techniques.

It is important for the student to realize that such four- and five-axis capabilities exist, but the three major axes to consider are X, Y, and Z. This text concentrates on these primary axes, beginning with X and Y programming examples and then Z axis capabilities.

INCREMENTAL SYSTEMS

Prior to any discussion involving incremental programming systems, the students must have a thorough understanding of incremental dimensioning. Carefully study the part in figure 4-5. The distance from the left edge of the part to hole 1 is 1.25. From hole 1 to hole 2, the distance is 1.50. The distance from hole 2 to hole 3 is 1.50, and 1.62 from hole 3 to hole 4. This is known as *incremental dimensioning*. It is also referred to as delta dimensioning. The word "delta" is derived from a Greek letter used to denote the difference between two quantities. In figure 4-5, each dimension is given incrementally from the last position to the next position.

An *incremental system* works according to the same principle; it positions the work or cutter in increments from the immediately preceding point. Calculations are made from the location of the tool or table to where it is going. The use of plus and minus signs involves a new aspect when used in the incremental mode. A positive X move does not refer to a specific rectangular quadrant, but directs the tool to move to the right along the X axis from its current position.

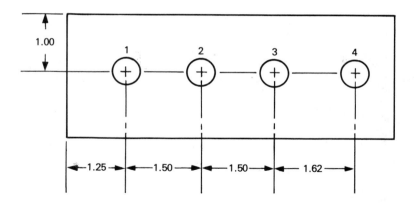

FIGURE 4-5
Incremental dimensioning

A negative X move directs the tool to the left. Similarly, a positive Y move positions the cutter up from the present location, and a negative Y is a command to move down. A positive Z directs the cutter away from the workpiece, while a negative Z is a move toward or into the workpiece.

ABSOLUTE SYSTEMS

A close examination of the workpiece in figure 4-6 will reveal its similarity to that shown in figure 4-5. The difference is the way the actual part is dimensioned. This type is known as absolute or baseline dimensioning because all positions must be given as distances from the same zero location or reference point. All dimensions are calculated from one zero point as indicated in figure 4-6.

An *absolute system* operates similar to absolute dimensioning. All positions are figured and punched in the tape relative to the same zero or reference point. All positional moves come from the same leading edge at all times, as opposed to an incremental system, where each succeeding move is an incremental distance from the last.

One advantage of absolute systems over incremental systems concerns positioning errors. If a positioning error occurs in an incremental system, all subsequent positions are affected and all remaining moves are incorrect. When a positioning error occurs in an absolute N/C system, a particular location is in error but subsequent positions are not affected. This is because all dimensions and each succeeding positional move are always based from the same zero or reference point. This is not to say, however, that the absolute system is superior to the incremental system. This philosophy originated in the early days of numerical control because purchasers were forced to choose between an absolute system and an incremental system. Consequently, all detailed workpiece drawings had to be dimensioned to conform with the

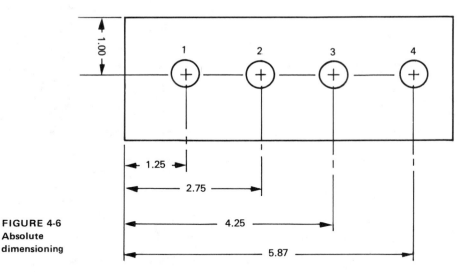

FIGURE 4-6
Absolute
dimensioning

particular control mode, or the programmer was forced to make the translation when preparing the program tape and manuscript.

Both absolute and incremental systems have their logical areas of application, and neither is always right or wrong. There are certain applications in
which both systems can be used most efficiently, sometimes even within the
same program. Most controls today are capable of working in either mode with
just a simple instructional code inserted to make the change. With the adapta-
bility of modern controls, the controversy over which is better is of little
importance. In many cases, the burden of decision is placed directly on the
programmer, who must have a thorough understanding of both modes and
be able to make the best use of each.

ZERO SHIFT SYSTEMS

Some type of *zero shift* capability exists on most N/C machines. This
capability implies shifting the zero location of the workpiece to any reason-
able location on the machine table. Machines and controls range from no
zero shift, or fixed zero to full range offset to full floating zero.

With fixed zero machines, there is no capability of changing the zero
location. The zero location is permanent and cannot be adjusted. Zero offset is used on fixed zero machines, and the control *remembers* the permanent
location of zero. A full floating zero machine has no fixed reference point
(zero point). For such a machine, the zero is established for each setup.

FIXED ZERO

Using a fixed zero system, the zero location on the machine table is
fixed; it cannot be moved or altered. This is illustrated in figure 4-7. The

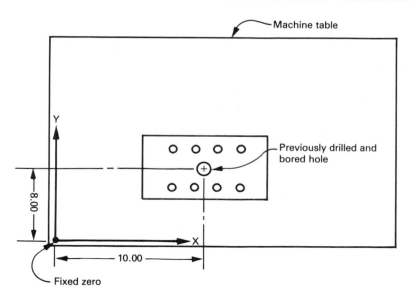

FIGURE 4-7
Fixed zero N/C system

operator is instructed by the programmer to set up the workpiece. The operator then dials in *X 10.0000* and *Y 8.0000* on the control decade switches and presses the cycle start button. The machine table moves so that the center of the spindle is positioned above the (10,8) point on the machine table. The operator must now clamp the workpiece on the table, push and tap the workpiece, and tighten and loosen the clamps until the previously machined hole "trams" in as centered in the spindle. This manual effort of adjusting clamps and moving the workpiece by small movements is required each time a setup is made with a fixed zero system. A small amount of shifting may occur with a fixed zero system. This amount can vary from manufacturer to manufacturer, but is generally only a few hundred thousandths of an inch.

FULL RANGE OFFSET

The machine zero in this type may be adjusted to any point on the machine, figure 4-8. Programmed dimensions must still be in the first quadrant, and all coordinate values must be positive. Most controls with the full range offset feature have no negative reading capability. The machine zero is still located in a permanent position, and the workpiece may be moved to any convenient location in the first quadrant.

FULL FLOATING ZERO

When using an N/C system equipped with *full floating zero,* the operator may locate the workpiece in any convenient location on the machine table. Once the workpiece is set up or positioned on the table, the operator then

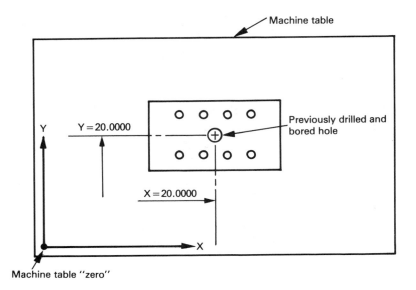

FIGURE 4-8
Full range offset N/C system

obtains the alignment positions for the particular workpiece program from the program manuscript. In figure 4-9, the alignment positions in X and Y are X = 0 and Y = 0 (0,0). These values are then dialed into the control. The operator depresses the cycle start button, and the machine rapid traverses to some location which, at this point, may not be relative to the workpiece setup just completed. The operator then uses the zero shift dials to change or move the zero location of the machine table to the setup location just completed.

Using the manual zero shift dials on the pendant station, shown earlier in figures 1-1 and 4-2, the operator changes or moves the machine table without changing the actual readout on the control for X and Y. The zero shift dials merely operate the table drive motors in X and Y, but their signal does not enter the memory section of the control unit. Turning these zero shift dials moves the table zero to the tram position for this particular workpiece.

Once the new workpiece is "zeroed" or "trammed" in, the zero shift dials are locked. The operator can then consistently run parts according to tape commands relative to the manuscript alignment position and the convenient zero location.

Full floating zero greatly enhances actual machine spindle cutting time by reducing setup time. The programmer and operator gain flexibility because the programmer now can make zero any place on the machine table, enabling positive and negative programming. The operator can leave high-production setups in place, and zero shift the machine to a new setup location on the machine table.

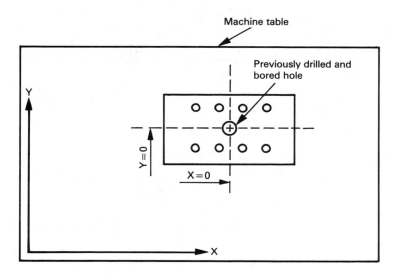

FIGURE 4-9
Full floating zero N/C system

REVIEW QUESTIONS

1. What is the importance of the Cartesian coordinate system to numerical control systems?
2. What two axes are normally considered when referring to a two-axis machine?
3. When programming for an N/C machine, is it necessary to write plus signs with all positive values? Why?
4. Who controls Z axis motion?
5. What is considered a positive Z movement? What is considered a negative Z movement?
6. Of the two N/C systems, absolute and incremental, which one would be most advantageous to you in a shop of your own? Why?
7. What is the relationship between absolute and incremental dimensioning of parts to absolute and incremental control systems?
8. What is a disadvantage of an incremental system that is an advantage of an absolute system?
9. Explain the difference between fixed zero, full range offset, and full floating zero shift.
10. What should a programmer know about absolute and incremental N/C systems?
11. On an N/C machine with a fixed zero, can a programmer write a manuscript using more than one quadrant? Explain your answer in detail.

CHAPTER 5

Tape Coding, Specifications, and Format

OBJECTIVES After studying this chapter, the student will be able to:

- Understand the importance of N/C tape specifications and standards.
- Distinguish between the different types of tape material available.
- Demonstrate an awareness of the different types of input media developed.
- Explain tape coding systems, their specifications and primary differences.
- Understand the physical processing of N/C tape.
- Discuss the different types of tape format in use today.
- Understand the fundamentals of the interchangeable or compatible tape format.

TAPE SPECIFICATIONS AND STANDARDS

A variety of control systems have been developed since the beginning of numerical control. Consequently, a wide variety of different methods for N/C data *input* has also been developed. In order to standardize this N/C information, an EIA subcommittee was formed to recommend a set of standards which would be acceptable to control system manufacturers and machine tool manufacturers and users.

Several types of *input media* were tried as numerical control evolved. The most common types were manual input, punched cards, magnetic tape, and punched tape.

Manual data input (MDI) is a means of telling the control system what to do through push buttons, dials, and switches. This system is completely operator-controlled and, therefore, is subject to human error. Speed and accuracy depend entirely on the operator's ability to find, interpret, and code in the correct information. Nearly all N/C machines today are capable of being operated in a manual mode even if just to position the X, Y, and Z slides.

Punched-card input was tried in some of the earlier N/C systems. All of the codes necessary to operate the machine tool were keypunched into a series of cards, and then run through the control system and decoded by "fingers," which activated electric circuits. Punched cards have been replaced as an input media for three main reasons. First, the card-reading device was much slower than a punched tape reader. Second, the individual cards were greatly affected by dirt and atmospheric changes. Third, the deck of punched cards was bulky and difficult to handle; the program could be scrambled if the cards were dropped.

Magnetic tape input was also tried with some early N/C applications. This consists of magnetic impulses located on a plastic tape. This tape is similar to that used on large-scale computers, and the stored information is formed in a manner similar to that of the punched-tape format. A magnetic tape the same length as a punched tape is capable of holding much more *data*. In addition, the tape is reusable once the magnetic information is erased.

There are four prime disadvantages of using magnetic tape. First, the tape reader and control system are more expensive than a punched tape system. Second, information on the tape can be accidentally erased if the tape is taken near a magnetic field. Third, the tape can become contaminated with metal particles and dust. Fourth, the information on the tape cannot be read to check for errors in programming.

Currently, manufacturers of numerical control systems commonly use the input form of punched tape. This method of communicating with the machine tool is economical and simple to prepare.

Punched tape has appeared in several different types of materials, sizes, and coding systems. The tape materials are primarily grouped under three main headings, although there are many combinations and variations. These will be discussed later.

The specifications for the size of punched tape have varied tremendously since the early years of N/C development. For example, the width of tapes has ranged from 1/2 inch to about 5 inches. Standardization became necessary in order to cut costs and establish common methods of programming between machine tool manufacturers. This was done through the cooperation of the EIA and the Aerospace Industries Association (AIA). Standardization covers two important categories: character coding, which will be discussed later, and physical dimensions.

As shown in figure 5-1, the physical dimensions of the tape have been standardized — one inch in width with eight tracks. It was determined that these tracks or *channels* were to run the length of the tape. The actual dimensions for thickness, hole spacing and size, and tolerances were also established at that time.

TAPE MATERIALS

The method of communication with an N/C machine tool through a punched tape input is quite economical and relatively easy to prepare.

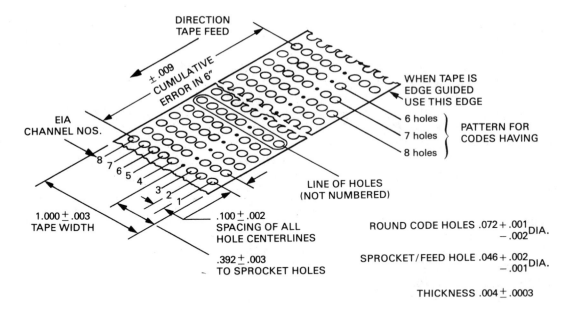

DIRECTION
TAPE FEED

±.009 CUMULATIVE ERROR IN 6"

EIA
CHANNEL NOS.

WHEN TAPE IS
EDGE GUIDED
USE THIS EDGE

6 holes
7 holes   PATTERN FOR
8 holes   CODES HAVING

8 7 6 5 4 3 2 1

LINE OF HOLES
(NOT NUMBERED)

1.000 ± .003
TAPE WIDTH

.100 ± .002
SPACING OF ALL
HOLE CENTERLINES

.392 ± .003
TO SPROCKET HOLES

ROUND CODE HOLES .072 + .001 DIA.
                          − .002

SPROCKET/FEED HOLE .046 + .002 DIA.
                             − .001

THICKNESS .004 ± .0003

FIGURE 5-1
Standard one-inch wide, eight-track tape with dimensions and tolerances shown

Several different types of materials have been used for punched tape. These materials are grouped in three categories: paper, Mylar (Du Pont Company's trade name for a tough plastic), and foil.

Paper tape is fairly inexpensive. It can be treated for oil and water resistancy. However, paper tape can be damaged, and photoelectric tape readers are somewhat sensitive to dirt, oil, and grease stains.

Mylar tape is more expensive; however, it is fairly indestructible and is not affected by oil, etc. Mylar tapes are sensitive to heat, and they will stretch when they become warm. However, when laminated with paper or aluminum, Mylar tapes are tougher and more durable for shop use. Many companies have standardized mylar tapes or the lamination of paper-mylar for use on their N/C machines in production.

Foil tape is a metallized material that is often used on high-production runs. High part production means that the tape is cycled continuously. Therefore, the metallized tape will wear better and last longer. In many situations, foil tape is not recommended because it is extremely hard on the tape preparation equipment punches.

Tapes can be purchased in other material combinations such as aluminum-mylar laminates, and new materials are constantly being tried and tested. Most of the N/C tapes, regardless of the material, can be bought in a variety of colors such as green, blue, yellow, and red.

TAPE CODING

Standardization was also necessary in tape coding. EIA assigned to each letter and number particular configurations of punched holes. These punched holes form codes across the width of the tape, and also use from five to eight tracks or longitudinal channels of holes. In numerical control work, eight-channel tape and two coding systems are used.

BINARY-CODED DECIMAL SYSTEM (BCD)

EIA adopted the *"binary-coded decimal"* (BCD) system for coding tapes. This is a system of number representation in which each decimal digit is represented by a group of binary digits forming a character. It incorporates the best features of both the binary and decimal systems, and gives a compact, easily simulated method of converting decimal dimensions into the analog voltage ratios which control the machine tool.

Binary code means base 2 as compared to our standard number system, base 10. In terms of electrical circuits, binary means that a signal is either ON or OFF. It is generally agreed that in specifying these signals, a zero (0) means "circuit off" and a one (1) means "circuit on." In an N/C tape with punched holes, the 0 is indicated by no hole and the 1 is indicated by a hole punched in the tape. With this limited amount of information, numerical control equipment must be able to send signals involving letters, numbers, and some special characters.

The binary system begins to the right with 0 or 1 and raises the digit 2 to a higher power as it moves to the left.

$$2^0 = 1 \text{ (zero power)}$$
$$2^1 = 2 \text{ (first power)}$$
$$2^2 = 4 \text{ (second power)}$$
$$2^3 = 8 \text{ (third power)}$$

The BCD system uses only the first four positions of the binary code (1, 2, 4, and 8), as well as some additional codes.

The decimal part of the system means that each number (0 through 9) can be written by the programmer and read by the machine control unit. The tape reader reads one digit at a time. The advanced electronic circuitry then places each number or digit in its proper decimal position. Figure 5-2 shows all character codes and the appropriate punches for one-inch-wide, eight-track tape (EIA RS-244).

Numbers 0 through 9 are formed in channels 1, 2, 3, 4, and 6 using the values determined by the binary system (base 2).

$$2^0 = 1 \text{ value of channel 1}$$
$$2^1 = 2 \text{ value of channel 2}$$
$$2^2 = 4 \text{ value of channel 3}$$
$$2^4 = 8 \text{ value of channel 4}$$

Channel 6 zero, a special binary code

Channel 5 odd-parity bit, used for checking

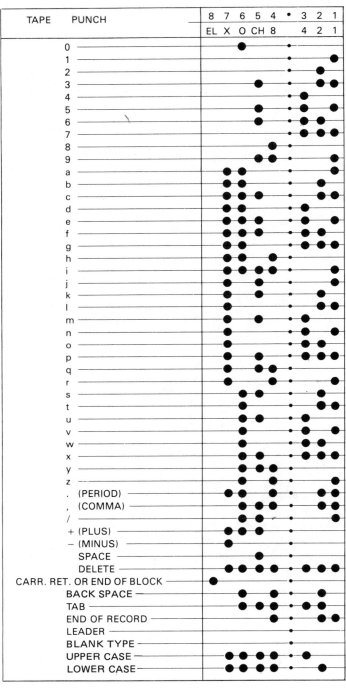

FIGURE 5-2
Character codes and punches for the BCD system (RS-244)

Method of coding letters		
Punch in CH. 6 & 7 plus numerical value	Punch in CH. 7 plus numerical value	Punch in CH. 6 plus numerical value
1 = A	1 = J	
2 = B	2 = K	2 = S
3 = C	3 = L	3 = T
4 = D	4 = M	4 = U
5 = E	5 = N	5 = V
6 = F	6 = O	6 = W
7 = G	7 = P	7 = X
8 = H	8 = Q	8 = Y
9 = I	9 = R	9 = Z

FIGURE 5-3
Letters are formed with the BCD system.

Alphabetical codes are formed using the number codes 1 through 9 in combination with channel 7 (X channel) and/or channel 6 (O channel). The alphabetical codes are also subject to the odd-parity check code in channel 5. Figure 5-3 shows how letters are formed with the BCD system. (Refer to figure 5-2 also.)

ASCII CODE

American Standard Code for Information Interchange (ASCII) is another coding system available on some N/C machines. This particular code was compiled by a committee from several different groups working with the United States of America Standards Institute. The overall objective of this group, now named American National Standards Institute (ANSI), is to obtain one coding system which will be an international standard for all information processing and communication systems.

There are several coding differences between BCD (EIA) and ASCII. ASCII provides coding for both uppercase and lowercase letters, while BCD codes are the same for both. The ten-digit codes (0 through 9) in ASCII are the same as BCD coding, but holes are punched in two additional tracks to identify the numbers and certain symbols. The ASCII letter codes, however, are quite different from those used in BCD. A comparison of the two systems is shown in figure 5-4.

Many of the more advanced control units now contain the necessary electronic circuits to handle both BCD and ASCII coding. A switch usually provides a choice of coding systems, or the control senses whether BCD or ASCII has been coded. Therefore, the machine is able to use either code.

BCD					(RS-244)			CHARACTER	ASCII				(RS-358)				
	•				•			0	•	•		•					
					•		•	1	•	•	•		•				•
					•		•	2	•	•	•		•			•	
		•			•		• •	3		•	•		•			•	•
				•	•			4	•	•	•		•	•			
		•		•	•		•	5		•	•		•	•			•
		•		•	•	•		6		•	•		•	•	•		
			•	•	•	•	•	7	•	•	•		•	•	•		
			•	•	•			8	•	•	•	•	•				
		•	•	•	•		•	9		•	•	•	•				•
•	•				•		•	a	•			•				•
•	•				•	•		b	•			•			•	
•	•	•			•	• •		c	•	•		•			•	•
•	•			•	•			d	•			•	•			
•	•	•		•	•		•	e	•	•		•	•			
•	•	•		•	•	•		f	•	•		•	•			
•	•	•	•	•	•	•		g	•			•	•			
•	•		•		•			h	•		•		•			
•	•	•	•		•		•	i	•	•		•	•			•
•		•		•	•		•	j	•	•		•	•		•	
•		•		•	•	•		k		•		•	•		•	•
•				•	•	• •		l	•	•		•	•	•		
•		•		•	•			m		•		•	•			
•		•		•	•		•	n		•		•	•			
•		•		•	•	•		o	•	•		•	•	•		
•		•		•	•	• •		p		•	•		•			
•		•		•	•			q	•	•		•	•			•
•			•		•		•	r	•	•		•	•		•	
	•	•		•		•		s		•		•	•		•	•
	•			•		• •		t	•	•		•	•	•		
	•	•		•	•		•	u		•		•	•			•
	•			•	•		•	v		•		•	•			
	•			•	•	•		w	•	•		•	•		•	•
	•			•	•	• •		x	•	•	•	•				
	•	•	•	•				y		•	•	•	•			•
	•		•	•			•	z		•	•	•	•		•	
•	•	•		•	•	•		.		•	•	•	•	•		
•	•	•		•	•	•		,	•	•	•	•	•			
•	•			•	•			/	•	•	•	•	•	•	•	
•	•	•		•				+		•	•	•	•		•	•
•				•				−		•	•	•	•	•		
		•		•				Space	•	•	•					
•	•	•	•	•	• •	•	•	Delete	• •	•	•	•	•	•	•	•
•				•				Car. Ret. or End of Block	•		•	•	•			•
	•		•	•		•		Back Space	Not Assigned							
	•	•	•	•	• •	•		Tab		•	•				•	
		•	•	•		• •		End of Record	Not Assigned							
				•				Leader	Not Assigned							
•	•	•	•	• •	•			Upper Case	Not Assigned							
•	•	•	•	• •		•		Lower Case	Not Assigned							

BCD - Binary Coded Decimal system
ISO - International Standards Organization
ASCII - American Standard Code for Information Interchange
The systems ISO and ASCII are the opposite of BCD in that they are even-parity coded.

FIGURE 5-4
Comparison of BCD and ASCII coding systems

TAB CODES

Tab codes are required by some N/C units to identify and arrange the information in the proper format. A *tab* is a nonprinting spacing action on tape preparation equipment. A tab code is used to separate words or groups of characters in the tab sequential format. The spacing action sets type-written information on a manuscript into tabular form. Tab codes can best be related to an office typewriter, on which the tab key returns the carriage to a preset position. In using this key, all characters will be lined up in columns. Tab codes are punched in N/C tapes to provide ease in reading a printout or *manuscript,* but they have no effect on machine action or movement.

PARITY BIT

Channel or track 5 on N/C tape, labeled CH, is used strictly as a *parity check.* This is a necessary function. Even though the equipment for tape reading and tape punching is extremely reliable, errors can still be made. EIA specifies an efficient method of correcting most mistakes; however, there is no system for identifying all the punching errors. Punching errors (partially punched, blocked, or unpunched holes) can cause mistakes in reading rows of information. Accuracy can be checked as follows: there must be an odd number of holes in every *row* for every character code punched in the tape. This is best illustrated in figure 5-2. For example, look at numbers 3, 5, 6, and 9 and at letters C, E, F, and I. The codes punched into the tape are correct without the parity check. However, when counting the number of holes across, you get an even number 2 or 4. The tape preparation equipment automatically punches a hole in track 5 to make an odd number of holes in that row. This is called an odd-parity system.

N/C systems are constantly checking for mispunched holes by sensing for an odd number of holes in each row. The parity check on N/C units is actually a safety device to help reduce the chances of error. When an even number of holes is detected in an N/C tape, the machine will stop automatically and the N/C system will indicate a tape error.

All systems do not have odd parity. The ASCII system uses an even-parity check; the N/C unit checks for an even number of holes in every row.

END OF BLOCK

The end of line (EL) or *end of block (EOB)* character indicates the end of a block of information, and is used primarily to indicate the end of an N/C instruction. It is also punched at the beginning and end of most N/C tapes. This is a punch in track 8 and is never combined with any other punches. The EOB code at the start of each program readies the parity check circuits and signals the control system that a program is about to start. This code at the end of a block of information signals the machine that it has received all the information necessary to perform that instruction, and the machine should satisfy that command. The EOB code is similar to a period at the end of a sentence. It signifies the end of one thought, command, or instruction.

Regardless of whether an N/C tape has been prepared by manual or computer methods, longer tapes have leader at the front and trailer at the rear of the coded section of tape. *Tape leader* is the portion of tape which contains only the sprocket holes. Leader is necessary if the machine tool must read the tape from a spool in the console reader. Enough leader is required on the spool for approximately two turns prior to reaching the first tape code. A trailer or *tape lagger* is the same as the leader except it trails after the program information. When the program has completely run through the tape reader, there should be approximately two turns on the spool to prevent a reading error and allow a tape rewind.

TAPE PREPARATION EQUIPMENT

When an N/C program or manuscript is completed, the next step is the actual punching of the tape. If the N/C program has been prepared by manual methods (listing all the codes and functions required for the machine to produce the workpiece), the N/C tape is punched by manual methods. This does not mean that someone punches each hole in the tape; however, the specific keys on a tape typewriter are depressed to type the manuscript and punch the tape. Several types of machines are available to prepare

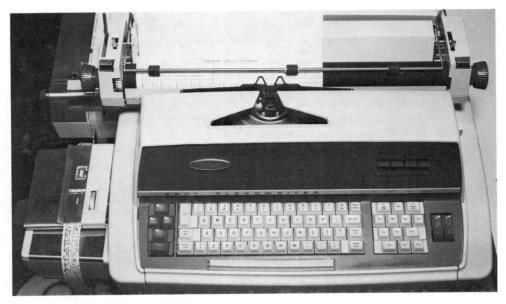

FIGURE 5-5
Manual tape preparation equipment (Courtesy of Cincinnati Milacron Inc.)

FIGURE 5-6
Another type of manual tape preparation equipment (Courtesy of Cincinnati Milacron Inc.)

N/C tape, as shown in figures 5-5 and 5-6. The electronic circuits allow the machines to punch the correct codes in the tape for every key on the tape typewriter. While the tape is being punched, a typed copy is made. These tape typewriters can duplicate a tape, type out the words or numbers from an N/C tape, and verify and correct tapes.

If the N/C program has been prepared by computer methods, the actual punching of the N/C tape is done through computer tape-punching equipment. An example of a computer tape-punching device which interfaces directly with a central computer is shown in figure 5-7. Since the program and tape are being generated by computer methods, there is no need for a typist to type the manuscript.

TYPES OF TAPE FORMAT

Tape format is the general sequence and arrangement of the coded information on a punched tape. This information conforms to EIA standards, and appears as *words* made of individual codes written in horizontal lines, as shown in figure 5-8. For example, there are five words that make up a block (one instruction) for this particular tape format. The most common type of tape format in current use is the word address or interchangeable format. However, some earlier control systems used the fixed sequential tab ignore and tab sequential format. There are still several of these types of control systems in use; therefore, they deserve some consideration in the discussion of tape formats. To illustrate the different types of tape formats,

FIGURE 5-7
**Computer tape-punching equipment (Courtesy
of Cincinnati Milacron Inc.)**

a five-word format consisting of sequence number, preparatory function, X dimension, Y dimension, and Z-axis depth selection will be used.

FIXED SEQUENTIAL FORMAT

Early generations of control systems used fixed sequential format. It contains only numerical data, arranged in a rigid sequence, with all codes necessary to control the machine appearing in every block. This format, illustrated in figure 5-9, has three main disadvantages. First, the typewritten copy of the tape is difficult to read because all the numbers appear as one long word. Second, the repetition of codes in every block means more work for the programmer and a longer tape. Third, no word address letter is used to identify the individual words.

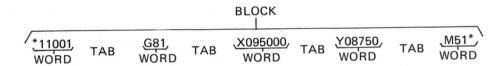

FIGURE 5-8
A block of information and the individual words

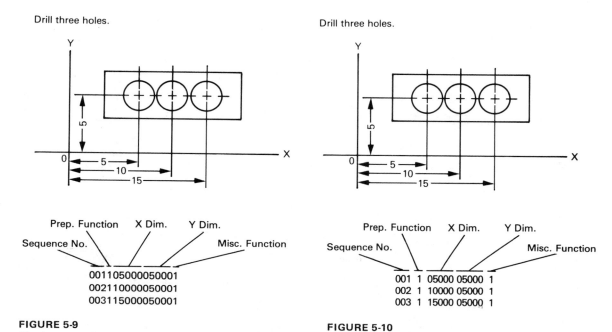

FIGURE 5-9
Fixed sequential format

FIGURE 5-10
Tab ignore format

TAB IGNORE FORMAT

This tape format employs the fixed sequential format with tab codes placed between each word. These codes make it easier to read the tape, and a neat printout is also produced. Tab codes cause no action in the control system. This feature gives rise to the name "tab ignore" format, figure 5-10, which contains unnecessary, repeated information and lacks the codes to make it more efficient.

TAB SEQUENTIAL FORMAT

The *tab sequential format* uses tab codes to a better advantage than does the tab ignore format. The tab code is not only used to separate words on a printout, but also to replace words that are unnecessarily repeated. As a result, the programmers' and typists' time and control reading time are reduced. Figure 5-11 illustrates the tab sequential format.

WORD ADDRESS FORMAT

The *word address format,* standardized by the EIA, uses a *letter address* to identify each separate word. By assigning an alphabetical code to each coordinate word and function word, the block format becomes more flexible. The purpose of a letter address is for word identification and to minimize the amount of data on a tape. Words do not have to appear in a rigid

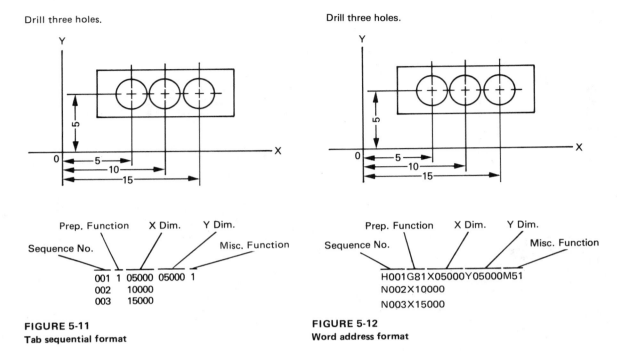

FIGURE 5-11
Tab sequential format

FIGURE 5-12
Word address format

format, and tapes are more interchangeable with machines in the same class. As seen in figure 5-12, the codes that change from one block to the next are the only ones that need to be programmed. Repeated codes can be omitted. Tab codes have been eliminated, and once again the printout of the program is relatively difficult to read.

INTERCHANGEABLE OR COMPATIBLE FORMAT

This format, which meets EIA standards, is probably the most sophisticated tape format in use today, although decimal point programming has recently proven successful and acceptable. The *interchangeable format* is basically the word address format, as illustrated in figure 5-13. This format permits the use of tab codes and the interchanging of words within a block of information. The length of the block is variable, as in the word address format. The tab ignore feature allows the tape to contain tabs because they are ignored by the control reader. As a result, a neat printout is produced.

It should be pointed out that even though the general tape format is followed, the specific tape format for a particular machine tool must also be followed. For example, an N/C lathe and an N/C machining center are two different N/C machines. Both are programmed using the interchangeable or compatible format, but the N/C lathe tape will not work in the machining center and vice versa. This is because each machine and control system has its own set of words that are recognizable. If the N/C control is fed unrecognizable

Drill three holes.

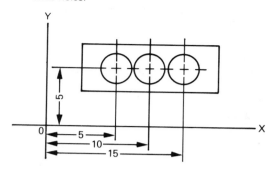

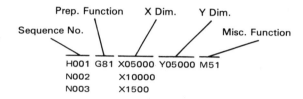

FIGURE 5-13
Interchangeable or compatible format

information, e.g., an N/C lathe tape accidentally loaded in a machining center control, a system failure will occur and the machine will not make any movements. Therefore, it is important to understand the particular tape format for the machine being programmed.

N/C TELLS A STORY

N/C does tell a story. It tells a story about the processing of a particular part on a particular machine. Essentially, much of the grammatical structure is the same when writing a story or a program. Look carefully at figure 5-14, and notice the similarities.

ENGLISH	N/C TERMS
STORY	*PROGRAM*
Sentence	Block
Word	Word
Letters	Letters & Numbers
Spacing	Tabs
Punctuation	End of Block

FIGURE 5-14
N/C tells a story

REVIEW QUESTIONS

1. Name the four types of input media, and briefly describe the advantages and disadvantages of each.
2. What were the two major categories covered under tape standardization? What two groups were responsible for the progress made in tape standardization?
3. Name the three types of tape materials, and discuss in general the advantages and disadvantages of each.
4. Explain the principle behind the binary-coded decimal system (BCD). Briefly explain how numerical and alphabetical codes are formed.
5. What are the basic differences between the BCD and ASCII coding systems?
6. What is the purpose of tab codes, and what effect do they have on an N/C unit?
7. What is the purpose of the parity check? What happens when a tape punch error is detected in the tape?
8. Why is the end of block (EOB) code punched at the beginning and end of every tape?
9. Explain the purpose of leader and trailer on an N/C tape.
10. What are the different methods of preparing N/C tape?
11. What are the five basic types of tape format? Which format is the most used and sophisticated?
12. Explain what is meant by tape format.

CHAPTER 6
Simple Part Programming — Conventions and Examples

OBJECTIVES After studying this chapter, the student will be able to:

- Understand basic part programming methodology as applied to a vertical N/C machine.
- Identify the various codes and functions in a typical block of N/C tape.
- Discuss related machine tool movements resulting from N/C coordinate information.
- Demonstrate a knowledge of various auxiliary function commands and their importance.

FUNCTIONS CONTROLLED BY N/C

Generally, an N/C program, regardless of machine type, consists of the heading, machine tape information, and operator information. The heading contains such clerical information as part name and number, drawing and fixture number, date, and programmer's name. Machine tape information contains all information necessary for proper machine operation, such as function codes. This information will vary, depending on the particular machine tool being used. The operator information section of an N/C program contains position number, depth of cut, operation description, and cutting tool information. Also included in this section is information on setup and machine alignment, as well as any other information the programmer feels the operator should know. This information is used strictly as an aid to the operator for a better understanding of the job.

The functions controlling N/C are part of the machine tape information, and are printed on a manuscript for operator and programmer reference. These functions consist of the following: sequence numbers; preparatory and miscellaneous functions; X, Y, and Z coordinate information; spindle speeds; feed rate; and depth selection. Each word consists of *alphanumeric codes,* which relate to a specific register in the machine control unit and cause an appropriate machine tool movement or action to occur.

This detailed information, in the form of letter and number codes, is best understood when applied to simple N/C machines like those in figure 6-1. Such a machine would have the following capabilities: a vertical, single-spindle machine with X and Y positioning and readout capabilities; full

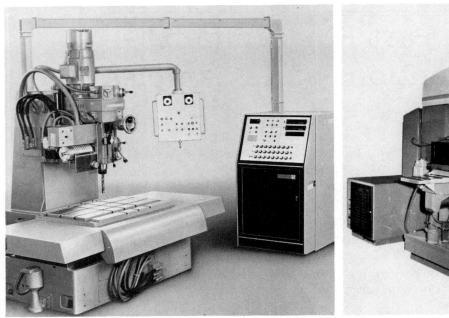
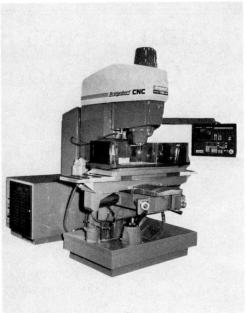

FIGURE 6-1
Simple N/C machine tools (A, Courtesy of Cincinnati Milacron Inc.; B, Courtesy of Bridgeport Machines)

floating zero; canned cycles for the various machining functions; and operator-controlled spindle speeds, feed rates, and depth selection.

SEQUENCE NUMBER

Normally, the first word in a block of information is the *sequence number.* This word, and others, appears as lighted numbers on the operator's console. Its primary purpose is to identify each block of information so that it can be distinguished from the rest and to indicate the block of tape being performed. The sequence number is usually a three-digit word but can occasionally be two or four digits if specified by a certain manufacturer. The sequence number is preceded by the letter H, O, or N, as shown in figure 6-2. Sequence numbers normally are progressive and informational rather than functional. The programmer may use the sequence number to indicate any nonfunctional information. The sequence number may be omitted if necessary. An example of this would be O001, N002, and N003.

SEQ. NO.	PREP. FUNC.	X POSITION	Y POSITION	F FEED RATE	CAM	MISC. FUNCT.
O---	G--	X+--- ----	Y+--- ----	F--- -	W--	M--

FIGURE 6-2
Typical tape format and word structure for a basic N/C machine

An H or O block should be used for the first block of information in every program; after every tool change; and for alignment and realignment blocks. Depending upon machine and control type, these rules are generally followed as full blocks of permanent information. They are frequently referred to through tape search. This feature, which is incorporated into most N/C units, allows the system to search the tape, through operator intervention, for a particular sequence number and stop when the number is found. Information may be temporarily bypassed and then recalled when needed.

The letter N is used for all other sequence blocks. It is not necessarily a full block of information and is, therefore, never used for searching.

X AND Y WORDS

Regardless of the type of N/C machine programmed, coordinate information is necessary for the machine tool to position itself. This coordinate information may be expressed using X, Y, and Z words. However, in this preliminary discussion, we will concentrate on X and Y words.

Coordinate input is normally a seven-digit number and the sign of the number, preceded by the letter X to indicate the X axis and the letter Y to indicate the Y-axis. The coordinate input normally occupies the third- and fourth-word positions in a tape format, as shown in figure 6-2. The X and Y words are written as X±******* and Y±*******. The position of the decimal point in coordinate information is also normally fixed in the tape format to allow four places to the right of the decimal point. The command of X+0043750 specifies the dimension of 4.3750 in the X axis and is accepted by the control system. Most controls have the capability of retaining the sign of the programmed word. Because of this, and depending upon the machine and control system, the sign of the number need only be programmed when it changes from the previous block of information. In an H or O block, the sign of the number should always be programmed to ensure the correct sign input when starting a series of operations.

It is important to remember when programming X and Y words that the actual centerline of the cutter is always programmed, as shown in figure 6-3. Establishing X and Y values for point-to-point (hole pattern) operations is easier because the direct X- and Y-word input consists of specific locations. Milling, in contrast, is more involved since the cutter radius must always be allowed for in programming for X and Y (centerline) locations.

Many N/C machines are equipped with manual input dials which allow the operator to enter specific X and Y coordinate information when needed. As a result, the machine can be moved

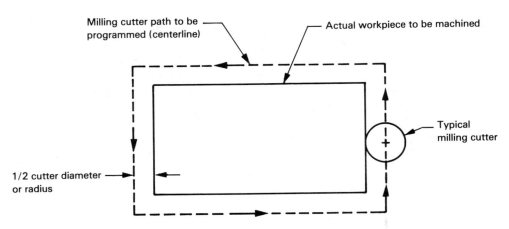

FIGURE 6-3
Importance of cutter radius compensation in calculating and programming centerline locations

to an operator-selected position in the manual mode of operation. This capability is extremely important and valuable in using X and Y coordinate input for setup, fixture, and workpiece alignment.

LEADING AND TRAILING ZERO SUPPRESSION

Decimal point positions in coordinate information are normally fixed to give four places to the right of the decimal point. Decimal points are not usually programmed. However, they can be in the more modern controls.

As mentioned, the machine actuation registers accept word-addressed information, and the machine tool responds to the corresponding signals. The coordinate information in the word-addressed X and Y registers enters, in most cases, in a right-to-left sequence, as shown in figure 6-4. This automatically positions the decimal point four places to the left of the X or Y word.

Study figure 6-4. While there are seven digit positions available, only five are needed. This means that the word could be written as X+0041250, and the two preceding zeroes could be programmed and punched into the tape. Either format would be acceptable to the control and machine. However,

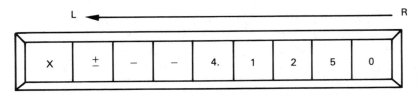

FIGURE 6-4
Fixed decimal location and right-to-left registration order

because the words enter the registers from right to left, the two *leading zeroes,* to the left of the *significant digits,* are insignificant and do not need to be programmed. Therefore, they can be suppressed. The zeroes are insignificant and have no effect on the programmed word. This omission of an insignificant digit is called *leading zero suppression.*

Some N/C systems have *trailing zero suppression* capabilities. This type of format works just the opposite of leading zero suppression. Machine actuation registers are filled from left to right rather than from right to left. The coordinate word X+41250, written with the leading zeroes suppressed, would be written as X+004125 with trailing zero suppression. The decimal point would be fixed, in this case, at three places to the right. The preceding two zeroes now become significant as the machine actuation registers fill from left to right.

PREPARATORY FUNCTIONS

The *preparatory function* or cycle code is a two-digit number preceded by the word address letter G (G**). This code, referred to as the *G code,* determines the mode of operation of the system. The preparatory function denotes some action of the X, Y, and/or Z axes. The X and Y axes will usually always position before the Z axis under conventional canned cycles.

A *canned cycle* is a combination of machine moves resulting in a particular machining function such as drilling, milling, boring, and tapping. A control with canned cycles may be more expensive than one without, but there is a definite gain to offset this cost. By programming one cycle code number, as many as seven distinct movements may occur. These seven movements would normally take at least six blocks of programming on a control without canned cycles. Using canned cycles, it is possible to realize a savings of up to 50% in programming time and up to one-third less data processing time. Tape length can also be reduced by at least one third with canned cycles. Most control manufacturers today have both canned and noncanned cycles as part of their standard control package.

Any further discussion of preparatory functions warrants the definition of two important terms.

1. *Rapid.* Positioning the cutter and workpiece into close proximity with one another at a high rate of travel speed, usually 150 to 400 inches per minute (IPM) before the cut is started.
2. *Feed.* The programmed or manually established rate of movement of the cutting tool into the workpiece for the required machining operation.

The following examples and descriptions of preparatory functions are some of the basic and more common canned cycle codes assigned by EIA. They are similar to the numbering systems used by many N/C machine and control manufacturers.

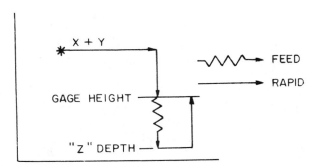

FIGURE 6-5
G81 — Drill cycle

Drill Cycle — G81. Figure 6-5 illustrates the G81 drill cycle. When a G81 cycle is programmed, the tool will:
1) rapid in X and/or Y.
2) rapid in the Z axis to *gage height*. (Gage height is the rapid distance the tool advances prior to contacting the part surface or the rapid distance the tool retracts after completing the cycle.)
3) feed in the Z axis to the Z depth.
4) rapid retract to gage height.
These four steps will occur in the same order every time a G81 cycle is called.

Dwell Cycle — G82. The G82 dwell cycle is illustrated in figure 6-6. When G82 is programmed, the tool will:
1) rapid in X and/or Y.
2) rapid in the Z axis to gage height.
3) feed in the Z axis to the Z depth.
4) dwell for the amount of time selected (usually .1 to 6 seconds).
5) rapid retract to gage height.
These five steps will occur in the same order every time a G82 cycle is called.

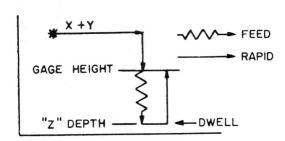

FIGURE 6-6
G82 — Dwell cycle

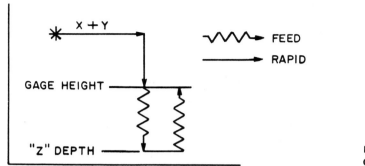

FIGURE 6-7
G85 — Bore cycle

Bore Cycle — G85. Figure 6-7 shows the G85 basic bore cycle. When the G85 is programmed, the tool will:
1) rapid in X and/or Y.
2) rapid in the Z axis to gage height.
3) feed in the Z axis to the Z depth.
4) feed retract to gage height.

These four steps will occur in the same order every time a G85 cycle is programmed.

Tap Cycle — G84. Figure 6-8 demonstrates the G84 tap cycle. When the G84 is programmed, the tool will:
1) rapid in X and/or Y.
2) rapid in the Z axis to gage height.

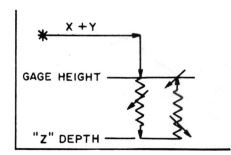

FIGURE 6-8
G84 — Tap cycle

3) feed in the Z axis to the Z depth.
4) reverse spindle direction and feed retract to gage height.
5) reverse spindle direction again at gage height.
These steps will occur in the same order every time a G84 cycle is called.

Basic Mill Cycle — G79. The G79 basic milling cycle is shown in figure 6-9. When the G79 is programmed, the tool will:
1) feed in X and/or Y.
2) rapid in the Z axis to gage height.
3) feed in the Z axis to the Z depth.
4) feed to following positions.
These four steps will occur in the same order whenever a G79 is programmed.

Cancel Cycle — G80. Figure 6-10 shows the G80 cancel cycle. When a G80 is programmed, the tool will:
1) rapid in X and/or Y.
2) cancel any Z motion.
These steps will occur in the same order every time a G80 cycle is called.

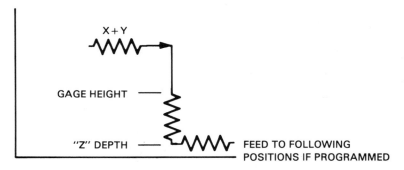

FIGURE 6-9
G79 — Basic mill cycle

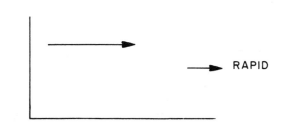

FIGURE 6-10
G80 — Cancel cycle

These codes illustrate some of the more common canned cycles currently in use. Additional preparatory functions will be introduced in later chapters.

FEED RATES

Feed rates govern the amount and rate of metal removal for a particular tool and type of workpiece material to be machined. Feed rates are normally measured in inches per minute (IPM) but can also be measured in inches per revolution (IPR).

For most machine tool and control manufacturers, the *feed rate* (F) word is coded directly in inches per minute. This is usually a four-digit number; the decimal point is assumed to be between the third and fourth digits (F***.*).

Example: .5 IPM F0005
30 IPM F0300
100 IPM F1000

The maximum and minimum feed rates per axis of the machine tool will vary depending on the machine and control manufacturer. These feed rates dictate the permissible feed range.

For some of the simpler tape formats in use, (figure 6-2), the feed rate (F) word is used only for milling operations. Feed rates for drilling, tapping, reaming, and boring are generally operator-controlled through machine adjustment and do not require tape input. On some of the more advanced machines and controls, feed rates govern rapid and feed movements for all cycles with operator override available.

SPINDLE SPEEDS

Spindle speeds (RPM) are not programmable for the type format discussed here. However, they are extremely important to tool life and the success of an N/C machine installation.

For the particular type of tape format displayed in figure 6-2, the *spindle speed* is totally operator-controlled as no S word may be either coded or direct RPM programmed. Spindle speed ranges, like feed rate ranges, will vary among machine and control manufacturers. Additional consideration will be given to spindle speeds in the chapters discussing N/C turning and machining centers.

DEPTH SELECTION

Other than tape-controlled Z-axis motion, operator-controlled Z-axis motion probably contains more variations among machine and control manufacturers than any other aspect of numerical control.

The method most commonly used with operator depth selection is to set manually several pairs of cams. Each pair of cams controls:

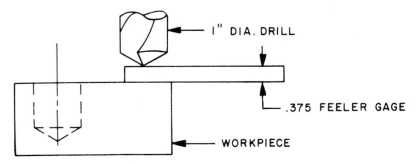

FIGURE 6-11
Rapid depth gage height being set using a .375 feeler gage

- the amount of rapid advance (downward motion) to the feed engagement point or gage height (rapid depth).
- the depth of travel at the operator preset feed rate to the depth required (feed depth).

In the format example in figure 6-2, the W word is the two-digit cam number word. Cam numbers generally range from W00 to W09. W00 selects manual operation and W01 through W09 selects cams 1 through 9 respectively. As already mentioned, each cam has two settings: rapid depth and feed depth. The rapid depth point may be set by using a feeler gage, as shown in figure 6-11. With the tool in this position and the console and cam selector properly set, the rapid ring may be rotated until the "on" cam light is illuminated. This means the limit switch is engaged and the rapid distance is set.

The feed depth may be established with the tool positioned on the workpiece surface, as shown in figure 6-12. If this method is used, the F (feed) ring should be set to compensate for the tool tip. In this case, with a 1.000 depth, the ring should be set at 1.000, plus an allowance for the drill point to ensure a one-inch deep hole. This drill point allowance is approximately .3 times the drill diameter for a 118° drill point. The remaining cams are set in the same manner.

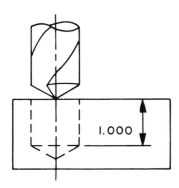

FIGURE 6-12
Feed depth being set with the tool positioned on the workpiece.

The W00 is a manual cam code which does not have a setting position. When programmed, the W00 code stops the reading cycle and allows the operator to perform manually the operation desired.

Several types of N/C machines use a rotating group of micrometer stops to control Z depth; others use electronically adjusted stops. In each case, for most of the simpler tape formats, the rapid and feed depths are operator-controlled, easy to set, and quite accurate.

MISCELLANEOUS FUNCTIONS

Miscellaneous functions perform a variety of auxiliary commands in numerical control. Generally, they are multiple-character, on/off codes that determine a function controlling the machine. These special features, functional at the beginning or end of a cycle, are two-digit numbers preceded by the letter M (M**). They activate *auxiliary functions* such as spindle start, coolant control, and program stop.

The following is a list of explanations of basic miscellaneous functions in accordance with EIA coding:

- *M00 (Program Stop).* This code *inhibits* the reading cycle after the movement or function has been completed in the block in which the program stop was coded. In addition, this code will also turn off the spindle and coolant if activated.

- *M01 (Optional Stop).* This code, like the (M00) program stops, inhibits the reading cycle after the movement or function has been completed in the block in which the optional stop was coded. This code will also turn off the spindle and coolant if activated. However, the code will only function if the operator has the control unit optional stop selector switch in the ON position. If the optional stop selector is in the OFF position, the (M01) will be read but no action stop will occur. For an automatic restart of either the spindle or coolant or both, it will be necessary to recode the proper miscellaneous function.

- *M02 (End of Program).* After the movement or function has been completed in the block in which the (M02) end of program was coded, this code will stop all *interpolation* (slide motion) and turn off the spindle and coolant. In addition to stopping the spindle and coolant and inhibiting any further slide motion, the (M02) end of program will rewind the tape to the leader (front) portion of the tape. The tape can also advance to the first H or O block if a *loop tape* is used, depending on which method is selected by the console switch. All registers are cleared when the M02 is read, and no information remains in the control unit's *memory.*

- *M06 (Tool Change).* This function should be coded in the last block of information in which a given tool is used. The specific machine tool design determines the sequence of events during the tool change. This code also stops the spindle and coolant, if activated, and retracts the tool to the full retract position.

- *M26 (Pseudo Tool Change).* Although this code will vary from one machine tool manufacturer to another, it is primarily used to generate a retraction from gage height to the tool change position. The M26 code will initiate the next tape command without stopping other than for reading time. This particular miscellaneous function is primarily used to avoid clamps and part obstructions.

For most machines and controls, the M00, M01, M02, M06, and M26 codes are effective only in the specific blocks coded. If they are to be used in two successive blocks, they must be repeated. It is generally not necessary to repeat other miscellaneous codes. In addition, the M02, M06, and M26 codes will generate full retract from the gage height only. Depending on the machine and control manufacturer, a suitable code (such as G81) must be used in the block with these codes to bring the tool to gage height.

Miscellaneous and preparatory functions are usually classified as either modal or nonmodal. *Modal* miscellaneous functions, such as M03 (spindle on clockwise), and preparatory functions, such as G81 (cycle drill), mean that these codes stay in effect until changed. These codes, as well as many others, remain operational regardless of how many succeeding blocks are programmed, until the code itself is changed. The modal condition can be changed or cancelled by programming a new miscellaneous or preparatory function.

Nonmodal codes, e.g., M00, M01, M02, M06, are effective only in the specific blocks programmed. As stated earlier, if they are to be used in two successive blocks, they must be repeated. Nonmodal codes, therefore, do not stay in effect until changed. They are operational only a block at a time.

SIMPLE PROGRAMMING EXAMPLES

In order to illustrate and explain practical applications of some basic N/C functions, specific examples have been detailed and illustrated. This method focuses attention on one particular element at a time and enables the student to understand and comprehend the process step by step.

In figure 6-13, sequence number O15, a drill cycle (G81) is used to move the table to position #1 from the previous position at a rapid rate. When position #1 is reached, the tool will rapid to gage height. At this point the tool will feed to depth. After reaching depth, the tool will rapid retract to gage height, and the next block of tape will be read (N16). Upon reading this block, the tool will rapid in X at gage height to position #2. When position #2 is reached, the tool will feed to depth. After reaching depth, the tool will rapid retract to gage height, and the M06 will retract the tool from gage height to the back or upper limit.

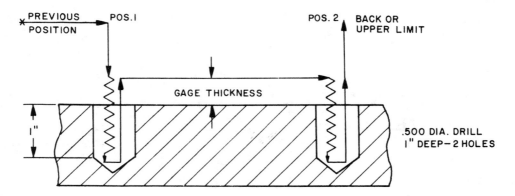

H/O or N SEQ. NO.	G PREP. FUNCT.	X POSITION	Y POSITION	I POSITION	J POSITION	F FEED RATE	ROTARY TABLE	ROTARY FEED RATE	W CAM	MISC. FUNCT.	POS. NO.
O15	G81	X+4 0000	Y+2 0000						W3		1
N16		X+6 0000								M06	2

FIGURE 6-13
G81 — Example of drill programming

NOTE: *The Z depth in the example was calculated as follows:*
  Z feed   Depth of cut   Drill point
          1.0000   (.3 × diameter of drill)
          1.0000   (.3 × .5000)
                  1.1500
The value of .3 is normally used for standard 118° drills.

In figure 6-14, sequence number O15, a drill dwell cycle (G82) is used to move the table to position #1 from the previous position at a rapid rate. When position #1 is reached, the tool will rapid to gage height. The tool will then feed to depth and dwell. After the dwell has terminated, the tool will rapid retract to gage height, and the next block of tape will be read (N16). Upon reading this block, the tool will rapid in X at gage height to position #2. When position #2 is reached, the tool will feed to depth and dwell. After the dwell has terminated, the tool will rapid retract to gage height, and the M06 will retract the tool from gage height to the back or upper limit.

NOTE: *There is generally a selector switch on controls which can be set to determine the amount of dwell. The setting is usually from .1 to 6 seconds.*

Study figure 6-15. Sequence number O15, a bore cycle (G85) is used to move the table to position #1 from the previous position at a rapid

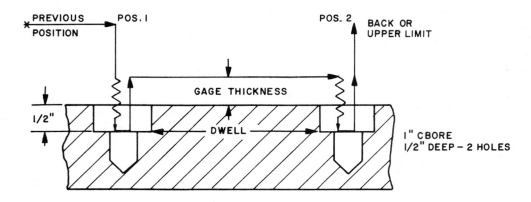

H/O or N SEQ. NO.	G PREP. FUNCT.	X POSITION	Y POSITION	I POSITION	J POSITION	F FEED RATE	ROTARY TABLE	ROTARY FEED RATE	W CAM	MISC. FUNCT.	POS. NO.
O15	G82	X+4 0000	Y+2 0000						W3		1
N16		X+6 0000								M06	2

FIGURE 6-14
G82 — Example of dwell programming

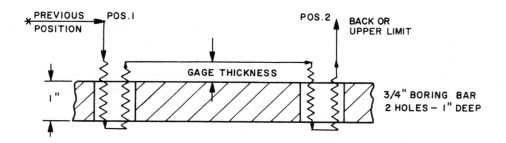

H/O or N SEQ. NO.	G PREP. FUNCT.	X POSITION	Y POSITION	I POSITION	J POSITION	F FEED RATE	ROTARY TABLE	ROTARY FEED RATE	W CAM	MISC. FUNCT.	POS. NO.
O15	G85	X+4 0000	Y+2 0000						W4		1
N16		X+6 0000								M06	2

FIGURE 6-15
G85 — Example of bore programming

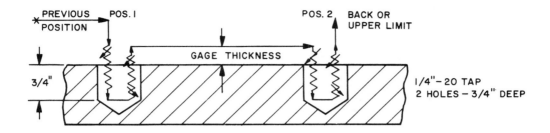

H/O or N SEQ. NO.	G PREP. FUNCT.	X POSITION	Y POSITION	I POSITION	J POSITION	F FEED RATE	ROTARY TABLE	ROTARY FEED RATE	W CAM	MISC. FUNCT.	POS. NO.
O15	G84	X+4 0000	Y+2 0000						W3		1
N16		X+6 0000								M06	2

FIGURE 6-16
G84 — Example of tap programming

rate. When position #1 is reached, the tool will rapid to gage height. At this point, the tool will feed to depth. After reaching depth, it will feed retract to gage height, and the next block of tape will be read (N16). Upon reading this block, the tool will rapid in X at gage height to position #2. When position #2 is reached, the tool will feed to depth. After reaching depth, it will feed retract to gage height, and M06 will retract the tool from gage height to the back or upper limit.

In figure 6-16, sequence number O15, a tap cycle (G84) is used to move the table to position #1 from the previous position at a rapid rate. When position #1 is reached, the tool will rapid to gage height. At this point, the tool will feed to depth. At depth, the spindle will reverse its direction and feed retract to gage height, and reverse the spindle direction again. The next block of tape will be read (N16). Upon reading this block, the tool will rapid in X at gage height to position #2. When position #2 is reached, it will feed to depth. At depth, the spindle will reverse and feed retract to gage height where the spindle direction will reverse again. The M06 will retract the tool from gage height to the back or upper limit.

In figure 6-17, sequence number O10, a G80 cancel cycle is used to position the table in X and/or Y at a rapid rate. The cam (W2) is stored in the control. In sequence N11, a G79 is used to rapid the tool to gage height and feed the tool to depth (.125 depth of cut). It is very important not to put an X and/or Y coordinate value in this block because the coordinate position would be satisfied first and then the Z motion would be satisfied. The sole purpose of this block is to feed the tool to depth. In sequence N12, the tool will move across the part at the programmed feed rate to position #2. In sequence N13, the G81 — M06 combination will retract the tool to the back or upper limit. A G80 along with an M02 or M26 could also be used for retraction.

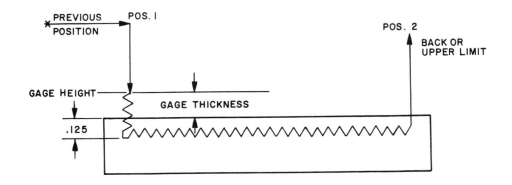

H/O or N SEQ. NO.	G PREP. FUNCT.	X POSITION	Y POSITION	I POSITION	J POSITION	F FEED RATE	ROTARY TABLE	ROTARY FEED RATE	W CAM	MISC. FUNCT	POS. NO.
O10	G80	X+10|0000	Y+10|0000			F5|0			W2		1
N11	G79										1
N12		X+13|0000									2
N13	G81									M06	2

FIGURE 6-17
G80 and G79 programming example

Figures 6-18 and 6-19 illustrate a complete and basic part and N/C program as applied to a vertical Cintimatic N/C machine. The routing for N/C processing the part would be to drill and tap complete along with milling the .25-inch deep step.

Examine the part in figure 6-18. There are two .500-inch holes drilled through, along with two .250-20-inch holes to be drilled and tapped completely through. In addition, a .25-inch deep step must be milled .38 inch wide across the entire width of the part. The material is 1018 steel.

The programmer, as discussed earlier, will determine the setup, alignment position, and coordinates to be used. In this particular case, the programmer has elected to use setup coordinate values of X = 10.0000 and Y = 10.0000. Consequently, all coordinate values will be positive and in the first quadrant, as illustrated in figure 6-20. This diagram indicates how, when calculating X and Y coordinates, actual part dimensions must be added to or subtracted from the alignment positions used by the programmer.

The following part program explanation is detailed on a block by block basis.

O001   .375 spot drill- spot first .250-20 hole- cam 1
       X position = 10.0000 + 6.5000 = 16.5000
       Y position = 10.0000 – 2.0000 = 8.0000

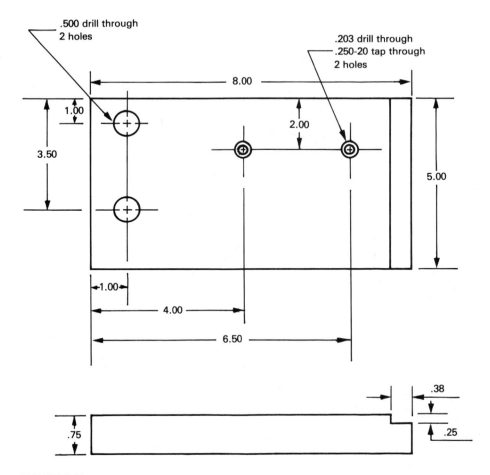

FIGURE 6-18
Mounting plate

N002   spot 2nd .250-20 hole X position = 10.0000 + 4.0000 = 14.0000
        same Y value (8.0000), does not need to be repeated
N003   spot 3rd hole- first .500 hole
        X position = 10.0000 + 1.0000 = 11.0000
        Y position = 10.0000 − 3.5000 = 6.5000
N004   spot 4th hole- 2nd .500 hole
        same X value (11.0000), does not need to be repeated
        Y position = 10.0000 − 1.0000 = 9.0000-tool change
O005   tool remains at same location (X = 110000, Y = 90000)
        new block of information- all values repeated-
        new cam-drill 1st .500 dia. hole

PART NAME: Mounting plate

MATERIAL: 1018 Steel

SETUP INFO: Locate part in vise — end stop on left end.
Set X = 10.0000 and Y = 10.0000.

SEQ. NO.	PREP. FUNC.	X POSITION	Y POSITION	FEED IPM	W CAM	MISC. FUNC.	RPM	DEPTH	REMARKS
O001	G81	X + 165000	Y + 80000	4.8	W01		800	.300	.375 spot drill
N002		X + 140000							
N003		X + 110000	Y + 65000						
N004			Y + 90000			M06			
O005	G81	X + 110000	Y + 90000	4.8	W02		600	.900	.500 drill
N006			Y + 65000			M06			
O007	G81	X + 140000	Y + 80000	5.6	W03		1400	.060	.203 drill
N008		X + 165000				M06			
O009	G84	X + 165000	Y + 80000	45	W04		1000	1.25	.250-20 tap
N010		X + 140000				M06			
O011	G80	X + 179950	Y + 45250	4.8	W05		400		.750 dia. two-flute end mill
N012	G79							.25	
N013			Y + 103750	F40					
N014	G80					M26			
N015						M02			

FIGURE 6-19
Mounting plate — basic N/C program

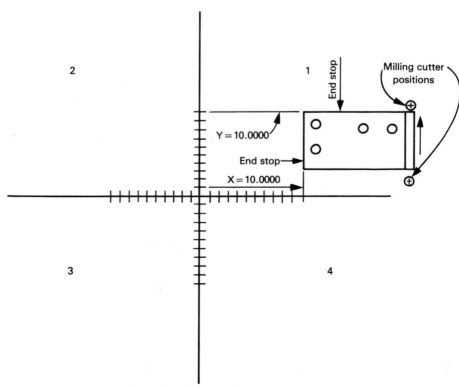

FIGURE 6-20
Diagram illustrating part to coordinate system location based on X and Y setup points as selected by programmer

N006   drill 2nd .500 dia. hole same X value (11.0000),
does not need to be repeated- Y value changes to
Y = 65000- tool change

O007   drill 1st .203 drill
X value (X + 140000) and Y value (Y + 80000) repeated
from center drill operation- new block of information-
all values repeated- new cam

N008   drill 2nd .203 hole
X value changes to X = 16.5000
Y value remains same at Y = 8.0000- tool change

O009   tool remains at same location (X + 165000, Y + 80000)
new block of information- all values repeated-
new cam- tap 1st .250 -20 hole

N010   tap second .250-20 hole
X value changes to X = 14.0000
Y value remains same at Y = 8.0000- tool change

O011   rapid traverse to position (G80) to begin milling step
coordinates calculated as follows: new cam

$$
\begin{array}{rl}
& \text{X} \\
10.0000 & - \text{X align position} \\
+\ \ 8.0000 & - \text{length of part} \\
\hline
18.0000 & \\
-\quad .38 & - \text{subtract step width} \\
\hline
17.6200 & \\
+\quad .375 & - \text{add half of .750 cutter dia.} \\
\hline
17.9950 & - \text{X coordinate value} \\
& \text{Y} \\
10.0000 & - \text{Y align position} \\
-\ \ 5.0000 & - \text{width of part} \\
\hline
5.0000 & \\
-\quad .375 & - \text{subtract half of .750 cutter dia.} \\
\hline
4.6250 & \\
-\quad .100 & - \text{subtract .100 additional clearance} \\
\hline
4.5250 & - \text{Y coordinate value}
\end{array}
$$

N012   G79 milling cycle actuated- quill rapids down to cam
depth at X + 179950 and Y + 45250 position

N013   .750 cutter feeds to end position at 4 IPM
machine feeds in Y direction- generally towards positive
stop- calculation is as follows (refer to figure 6-20):

$$
\begin{array}{rl}
& \text{Y} \\
10.0000 & - \text{Y position} \\
+\quad .375 & - \text{half of .750 cutter dia.} \\
\hline
10.375 & - \text{ending Y coordinate value}
\end{array}
$$

N014   cancel G79 milling cycle (G80)
retract quill (M26)

N015   end program (M02)

CINCINNATI ACRAMATIC SERIES 5 225

PART NAME | DRAWING NO. | REVISION | FIXTURE NO. | PROGRAMMED BY | PART NO.
PAGE OF PAGE | DATE

SET UP AND TOOL INFORMATION

H/O or N SEQ. NO.	G PREP. FUNCT.	X POSITION	Y POSITION	I POSITION	J POSITION	F FEED RATE	ROTARY TABLE	ROTARY FEED RATE	W CAM	MISC. FUNCT.	POS. NO.	SPD. FD.	RPM	DEPTH OF CUT	REMARKS
N81		X+0	Y+6 0000	I+0	J+0						P5				
N82		X+6 0000	Y+0	I+0	J+0						P2				
N83	G81									M26	P2				
O84	G80	X+5 2774	Y-3 6953			F16 0			W1		P14	8	1500	.3125	2" END MILL
N85	G79										P14				
N86		X+5 2542	Y-3 6791								P14				
N87		X+3 6791	Y+5 2542								P15				
N88		X-5 2542	Y+3 6791								P16				
N89		X-3 6791	Y-5 2542								P17				
N90		X+5 2542	Y-3 6791								P14				
N91	G81									M06	P14				
O92	G81	X+4 3750	Y-4 3750			F15 0			W5		P18	15	1500	1.125	59/64 DRILL
N93		X-4 3750	Y+4 3750								P20				
N94		X+4 3750									P19				
N95		X-4 3750	Y-4 3750							M06	P21				
O96	G84	X-4 3750	Y-4 3750			F15 0			W6		P21	15	220	Approx.	3/4-14 PIPE TAP
N97		X+4 3750	Y+4 3750							M06	P19				
O98	G85	X-4 3750	Y+4 3750			F10 0			W7		P20	10	2100	.500	1" CARB. BORE
N99		X+4 3750	Y-4 3750							M06	P18				
O100	G81	X+2 5000	Y-2 5000			F15 0			W8		P22	15	2100	.875	5/16 DRILL
N101		X-2 5000	Y+2 5000							M06	P23				
O102	G84	X-2 5000	Y+2 5000			F21 0			W9		P23	21	350	.375	3/8-16 N.C. TAP
N103		X+2 5000	Y-2 5000								P22				
N104	G80	X+0	Y+8 0000							M02	P24				UNLOAD

FIGURE 6-21
Full-page manuscript form (Courtesy of Cincinnati Milacron Inc.)

As was discussed earlier, these examples represent only some of the more widely used basic codes and functions. Most N/C machines have more capabilities than those discussed here. Some of the other capabilities will be discussed in detail later.

A full-page manuscript form is illustrated in figure 6-21. This is typical of a *manual part programming* form for the type of machine and format discussed in this chapter. Before actual programming begins, all information on speeds, feeds, hole locations, and so on must be assembled in an orderly arrangement to simplify the process and reduce errors. Forms similar to that in figure 6-21 help organize all necessary information on machining sequence, tool numbers and description, feeds and speeds, etc. Having this information tabulated simplifies manual and computer methods of part programming.

REVIEW QUESTIONS

1. Name and describe briefly the three main parts of an N/C program.
2. What is the primary purpose of the sequence number in an N/C program? What are the general rules determining when a O block should be used?
3. How is the position of the decimal point accounted for when coding N/C coordinate information?
4. What important cutter-to-workpiece relationship must be considered when programming an N/C milling operation?
5. Explain leading zero suppression and its significance to loading the machine actuation registers.
6. In general, what is meant by a preparatory function?
7. What is the difference between canned cycle and noncanned cycle preparatory functions?
8. Name and describe some advantages of canned cycle preparatory functions.
9. What is the basic difference between the G81 and G82 codes?
10. What occurs each time the G84 tap cycle is programmed?
11. Describe in detail the importance of programmed feed rates in an N/C program?
12. Briefly explain how operator-controlled cams function.
13. What are miscellaneous functions? What type of machine commands are controlled by these codes?
14. What types of information should be organized and completed prior to beginning an N/C program?

CHAPTER 7

Other Functions Controlled by N/C

OBJECTIVES
After studying this chapter, the student will be able to:

- Understand linear and circular interpolation and their functional operations.
- Explain the importance of preset tooling and tool length compensation.
- Describe how work surfaces and related changes are programmed.
- Discuss the primary differences between random and sequential tooling.
- Understand how varying differences in cutter diameters are compensated and programmed.
- Demonstrate a knowledge of additional tooling functions and their importance.

LINEAR INTERPOLATION

Linear interpolation is programmed points connected by a series of straight line movements. These straight line movements will result in the desired contour when programmed correctly and in sufficient supply.

The best way to represent linear interpolation and related moves is to first review figure 7-1, a circle with a hexagon inside. Each of the sides of the hexagon is a chord or straight line segment connecting extremities of an arc. If more chords were added inside the circle, the chords together would come closer to being a perfect circle. The number of straight line segments required for machining is determined by the maximum tolerance allowed between the design of the contour and the programmed line segment. The accuracy, then, of the arc or contour to be machined will depend on the size of the line segments and the number of programmed points.

Figure 7-2 illustrates the relationship between the desired cutter path and the actual cutter path along with programmed points for one linear interpolation move. A programmer using linear interpolation to program a circular path must determine the maximum allowable part error and center-line error. As the acceptable part error gets smaller, more programmed points are required to generate the curve. In figure 7-2, only two programmed

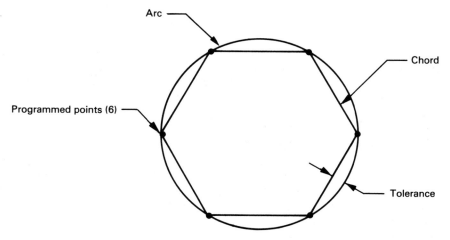

FIGURE 7-1
A circle with hexagon inside illustrating a chord-to-circle relationship

points are shown, but all coordinate positions necessary to complete the curve with an acceptable surface finish must appear in the program. This means the more programmed points that are needed to define the line segments, the greater the length of the N/C program. It also means that the amount of programming time increases as the number of line segments increases because each point, regardless of the distance between them, must be programmed.

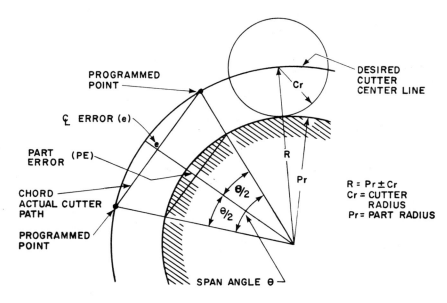

FIGURE 7-2
Diagram depicting the relationship between the desired cutter path and the actual cutter path

Linear interpolation moves to generate the desired contour are controlled by the rate of travel in two directions proportional to the distance moved. A linear interpolation cut requires the X and Y drive motors to run at unequal speeds, thus dictating an elaborate control/drive system. Most of the more advanced controls contain sophisticated control/drive systems and are capable of linear interpolation moves.

CIRCULAR INTERPOLATION

Circular interpolation allows the programmer to move the cutting tool in a circular path ranging from a small arc segment to a full 360-degree *span*. The *cutter path* along the arc is generated by the control system. Arcs up to 90 degrees can be handled by the control in one block of information. On some control systems the programmer will have to program four blocks of information to obtain a 360-degree arc. The total number of blocks programmed normally will vary depending on the specific type of circular interpolation used.

There are different versions of circular interpolation available on modern equipment, along with different programming techniques. On some MCUs, circular interpolation is limited to the XY, XZ, or YZ axes; the control cannot simultaneously interpolate circular movements for all three axes. This text will concentrate on the concept of circular interpolation and a specific application.

The four basic points necessary to program this particular type of circular interpolation are shown in figure 7-3. The points are:

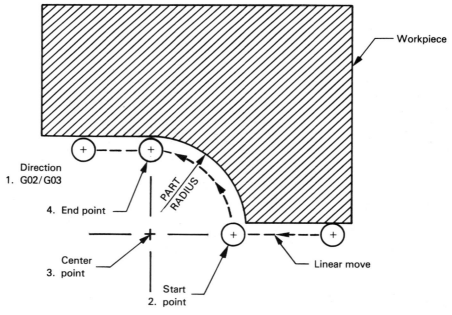

FIGURE 7-3
The four basic elements of circular interpolation

- preparatory function (G02/G03)
- start point
- center point
- end point

The preparatory function codes G02 and G03 are EIA standard codes. They are used for programming circular interpolation. These codes determine the direction of the circular path as viewed from the positive end of the axis that is perpendicular to the plane of interpolation (3:00 position in X and Y).

G02 = circular interpolation clockwise (CW)
G03 = circular interpolation counterclockwise (CCW)

These codes are programmed in the block where circular interpolation becomes effective. They remain effective until a new preparatory function code is programmed.

The start point, an X, Y, and/or Z coordinate, is usually the result of a previous arc (circular interpolation) or the end point of a line (linear interpolation). The start point is always described by X, Y, and/or Z words, and it normally positions the cutting tool for the following circular move.

The center point (an X, Y, and/or Z coordinate) is the center of the circular arc. The center point is described by I, J, and K words. The I word describes the X coordinate value; the J word describes the Y coordinate value; and the K word describes the Z coordinate value. Usually, the I, J, and K words are absolute values regardless if they are programmed in the absolute or incremental mode. The center point is modal in the absolute mode but must be programmed in each block in the incremental mode.

The *end point,* and X, Y, and/or Z coordinate, is the final point where the centerline of the cutter path completes the circular arc. The end point is described by X, Y, and/or Z words, and must be programmed in every block using circular interpolation. When programming circular interpolation, arcs using more than one 90-degree quadrant, the point where the arc crosses into another quadrant must be programmed as an end point. The control assumes this end point to be the start point for the next 90-degree circular span. In the next block of information, it is necessary only to program a new end point. A new center point is programmed only if the center coordinates of the arc are changed. If the end point does not fall on the arc defined by the center and the calculated radius, the cutter path and its rate are unpredictable as no program error will be indicated in most cases.

The following program illustrates the use of circular interpolation for the part depicted in figure 7-4.

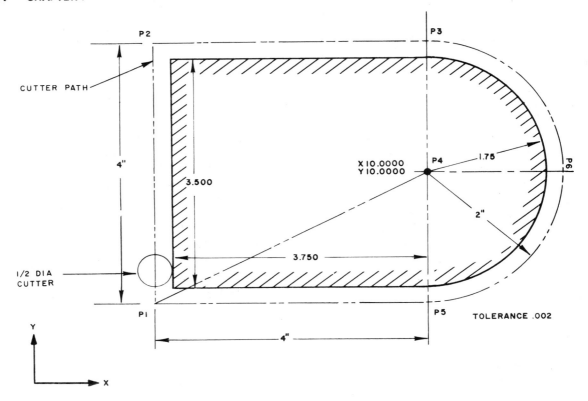

FIGURE 7-4
Part programmed with linear and circular interpolation

H001	G80	X + 060000 Y + 080000			P1
N002	G79	DEPTH			P1
N003		Y + 120000			P2
N004		X + 100000			P3
N005	G02	X + 120000 Y + 100000	I + 100000	J + 100000	P6
N006		X + 100000 Y + 080000	I + 100000	J + 100000	P5
N007	G79	X + 060000			P1
N008		FULL RETRACT			P1
N009	G80	X + 100000 Y + 100000		M02	P4

NOTE: *(N005 and N006) The end point of the first arc becomes the start point of the second arc.*

If the part in figure 7-4 was programmed with linear interpolation, it would require approximately thirty-six linear spans to program the 180-degree arc. The programming of the 180-degree arc using circular interpolation requires only two blocks of information to be programmed for two

90-degree arcs. Additionally, the tape length for this program would be approximately 1.5 feet as compared to a tape length of approximately 6.3 feet for the linear program.

There are other types of interpolation available with modern MCUs, including parabolic, cubic, and helical interpolation. *Parabolic interpolation* is used to approximate curved sections that conform to either a complete parabola or a portion of one. Cubic interpolation is applicable to automotive shapes requiring third-degree curve interpolation of sheet metal forming dies. Helical interpolation lends itself to helical cutting applications where the control must calculate the radius of the helix from the start of the arc to the center. If the end point does not fall on the arc defined, the control will interpolate a helical arc as far as possible and then move to the programmed end point with a linear move.

Parabolic, cubic, and helical interpolation are specialized applications for the particular needs of industries that manufacture components with complex shapes. The most common interpolation routine is circular. It lends itself to a variety of common manufacturing applications.

PROGRAMMABLE Z DEPTH

Most modern machine tools have programmable Z motion. Although the procedures differ in some aspects between manufacturers, the Z motion must be programmed accurately by the programmer if the machine tool is to produce quality workpieces.

Generally, the Z word is a seven-digit word preceded by a plus or minus sign and the Z-word *address.* It controls the depth to which a particular tool enters the workpiece for each machining operation. The Z motion is relative to a rapid to-position move of some type, through either an R word (to be discussed later) or a rapid traverse fixed rate with a specific Z.

If only the Z word is programmed to control the entire Z axis movement, the following formula can be used in most cases:

$$Z = PS + CL + TL$$

where: Z = distance from ZO to spindle gage line
  PS = distance from ZO to the part surface
  CL = clearance (if needed)
  $*TL$ = tool set length from spindle gage line to cutting edge

To find the Z value for the part and tool in figure 7-5, the following calculation is made:

---

*Since many systems are arranged with a tool length storage feature, the control in some cases will add the tool set length (TL) to the programmed Z value. In this case, the programmer need not include the tool set length dimension when calculating the Z word.

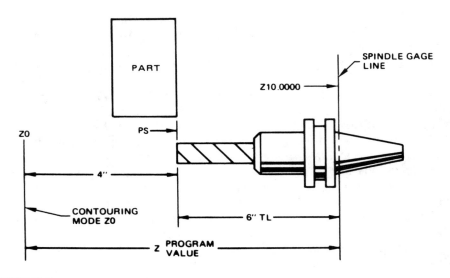

FIGURE 7-5
Z value calculation, showing part and tool

$$Z = PS + CL + TL$$

Where: PS = 4.0000
TL = 6.0000
CL = 0.0000 (cutter positional to part surface)

Z = 4 + 0 + 6
Z = 10.0000

Additional applications of Z motion will be discussed later in a section on R work plane as applied to fixed cycle programmable Z.

TOOL LENGTH COMPENSATION

Allowances must be made for differences in tool lengths because they may vary considerably from one tool to the next. Most companies overcome these differences in tool length by having all tools preset. Presetting tools establishes consistency of length each time that particular tool is required.

Some companies have prepared *tool assembly drawings* for each tool. These tool layouts are prepared to describe the cutting tool and its setting length, figure 7-6. Each tool is given a symbolic number. On machines and controls equipped with tool assembly number storage, this is the number accessed on the machining center and stored within the control.

The standard center drill for the tool assembly shown in figure 7-6 must be set with dimensions consistent as described on the drawing. Generally, individuals in tool stores, a tool preset area, or the machine operator will assemble the drill in its holder according to the tool assembly drawing each

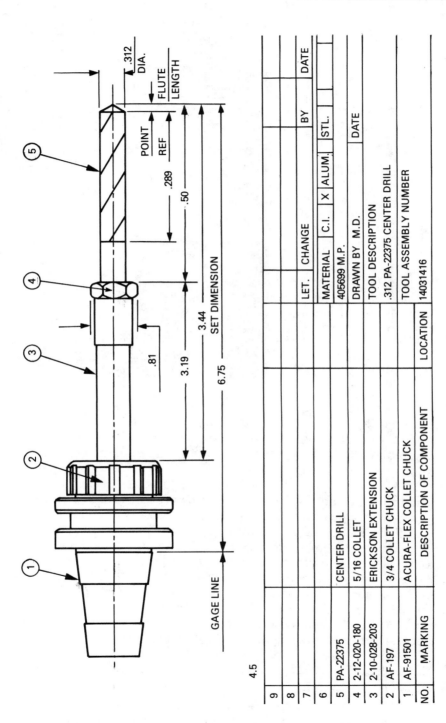

NO.	MARKING	DESCRIPTION OF COMPONENT					
9							
8							
7							
6							
5	PA-22375	CENTER DRILL					
4	2-12-020-180	5/16 COLLET					
3	2-10-028-203	ERICKSON EXTENSION					
2	AF-197	3/4 COLLET CHUCK					
1	AF-91501	ACURA-FLEX COLLET CHUCK					

LET.	CHANGE	C.I.	X	ALUM.	STL.	BY	DATE
MATERIAL 405699 M.P.							
DRAWN BY M.D.				DATE			
TOOL DESCRIPTION							
.312 PA-22375 CENTER DRILL							
TOOL ASSEMBLY NUMBER							
LOCATION	14031416						

FIGURE 7-6
Typical tool assembly drawing

FIGURE 7-7
An electronic preset tool gage
(Courtesy of Cincinnati Mila-
cron Inc.)

time the tool is required. Preset and tool store personnel will sometimes maintain a duplicate set of drawings consistent with tools selected for use in N/C programming. When the part is ready to be processed, the preset area assembles all the tools specified by the part programmer according to each tool assembly number.

In some cases, an electronic tool gage is used to obtain the overall length of the tool. An example of this type of tool preset gage is shown in figure 7-7. Recent advances in tool presetting have allowed electronic tool gages to be interfaced directly to the MCU for direct tool length entry into the control system. This type of gaging and machine control interface helps obtain accurate tool lengths and reduce human error.

It is important to remember that the machine must be made to cut metal as constant as possible to maximize productivity. Therefore, arrangements must be made to replace dull cutting tools in a minimum amount of time and still maintain accurate tool lengths. Additional consideration of tool lengths and tool length compensation will be covered in Chapter 9.

R WORK PLANE

The R word in N/C programming refers to the work surface or rapid distance being programmed. The Z rapid position, as the work plane is

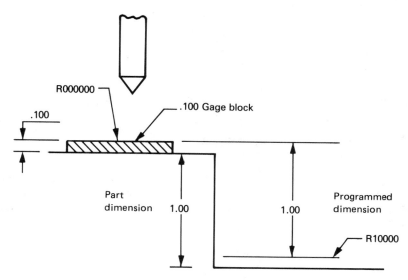

FIGURE 7-8
Establishing and calculating R000000

sometimes called, is established as the R work surface location at a specified height on the part. All other work surfaces are relative to this location and thus height or distance is established. On some N/C machines, this is the highest part surface and will be used in our example.

As can be seen in figure 7-8, R000000 is sometimes established at the highest work surface by setting all tools at this surface. When setting up the job, the operator places a .100 gage block on top of the highest work surface (or any other location the programmer has specified). The turret is lowered, or the spindle is brought up to the workpiece, by jog control, so that the tool in the spindle just touches this .100 gage block. The operator then sets the tool length compensation switches if applicable, and the setting to R000000 is complete for that tool. This same process is repeated for all tools as they are all random length at the same height. The gage height distance is normally .100. Generally, it is not necessary to add in this distance when changing work surfaces, as can be seen from our example. In most cases, the gage height distance of .100 is built into the MCU. Whenever a Z feed motion is called for, the .100 will be automatically added to the programmed Z depth.

When programming the R word, the programmer must always know the location of each tool's tip, particularly when changing work surfaces. When changing work surfaces, the programmer simply programs the R or rapid distance as indicated on the blueprint. As shown in the example, a rapid distance of one inch is specified as an R10000 and will position the tool .100 above the part surface.

The programmed Z depth of .375 (Z3750) at R000000 could also be programmed at an R10000. In addition, a programmed R is modal. That is,

O/N SEQ.	G PREP. FUNCT.	X± POSITION	Y± POSITION	Z± POSITION	R± POSITION	I/J/K POSITION	A/B/C P/Q ± WORD POSITION	F/E FEED RATE	S SPINDLE SPEED	D/H WORD	T TOOL WORD	M MISC. FUNCT.
Ø 5	G81	X+ 80000	Y+ 100000	Z - 10000	R+100000		B 0	F 150	S 720		T3	M03
N 6			Y+ 140000		R 90000							

— In Sequence Number Ø5, the G81 code rapid advances the X and/or Y axes simultaneously to Pos. 1 from the previous position. When Pos. 1 is reached, the Z axis will rapid to the R 10.0000 plane, and will feed to the programmed depth of 1″ at the programmed rate. After reaching depth, the Z axis will rapid retract to the R 10.0000 plane, and the next block of information will be read and acted upon.

— In Sequence N6, the G81 code rapid advances the Y axis — at the R 10.0000 plane — to Pos. 2. Then the Z axis will rapid to the new R plane (R 9.0000), and feed to the programmed depth of 1″. After reaching depth, the Z axis will rapid retract to the R 9.0000 plane.

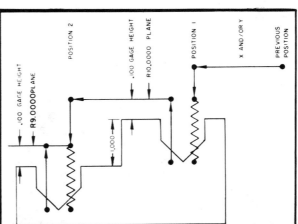

FIGURE 7-9
Changing work plane — high to low

O/N SEQ.	G PREP. FUNCT.	X ± POSITION	Y ± POSITION	Z ± POSITION	R ± POSITION	I/J/K POSITION	A/B/C P/Q POSITION ± WORD	F/E FEED RATE	S SPINDLE SPEED	D/H WORD	T TOOL WORD	M MISC. FUNCT.
N 4	G81	X+ 80000	Y+ 90000	Z- 10000	R+ 90000		B 0	F 150	S 720		T3	M03
N 5	G80				R 100000							
N 6	G81		Y 60000									

— In Sequence Number Ø4, the G81 code rapid advances the X and/or Y axes simultaneously to Pos. 1 from the previous position. When Pos. 1 is reached, the Z axis will rapid to the R 9.0000 plane, and will feed to the programmed depth of 1" at the programmed rate. After reaching depth, the Z axis will retract to the R 9.0000 plane, and the next block of information will be read and acted upon.

— Sequence Number N5, with the G80 code, will rapid retract the tool from the R 9.0000 plane to the R 10.0000 plane.

— Sequence No. N6, with the G81 code re-programmed, will rapid advance the Y axis to Pos. 2 and will drill a hole 1" deep on the higher level.

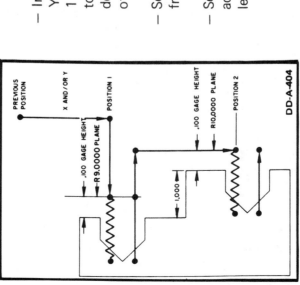

DD-A-404

FIGURE 7-10
Changing work plane — low to high

the programmed R plane or work surface stops or locks in the MCU until a new R word is programmed.

When a new R word is programmed, a change in work surface will result. Whether the programmer wants to move the tool to a higher or lower work surface is of considerable importance. Machine tools that require an R word to be programmed usually have a specific order of processing blocks of information. For a block of information which contains an X, Y, R, and Z word, the X and Y words have the highest priority and will be satisfied first. The R word has the second highest priority and will be satisfied second. The machine tool will respond with a rapid traverse move to satisfy these words at the rapid traverse rate of the machine tool (sometimes 400 IPM). The Z word has the lowest priority, and the tool will feed to depth only after the X and Y words have positioned the cutting tool and the R word has satisfied the programmed work surface. No problems are presented when changing from a higher to lower work surface, figure 7-9, because of the order in which the priorities are followed — X and Y first, R second, and Z feed last.

When changing from a lower to a higher work surface, the priorities are the same but must be programmed differently. Programming a block of information with X, Y, R, and Z in the same line of information, when moving from a lower to higher work surface, will result in a collision of the tool at rapid traverse against the side of the workpiece. This could cause tool breakage, injury to the operator, and considerable damage to the toolholder and machine spindle, as well as other serious effects. This problem must be overcome by retracting the tool to the higher work surface first with a G80, then reinstating the desired code plus the new X and Y positions, as shown in figure 7-10. Being aware of clamp locations when programming is also important. The programmer may find it necessary to avoid accidents by programming Z avoidances in order to move over or around clamps.

When preparing a program, the programmer must know on what surface the programmed tool is, where the tool will rapid to next, and the path the tool will follow to reach the next position. Herein lies a serious problem when programming N/C equipment. Programmers often overlook these potential collisions when checking their manuscripts, and critical and expensive accidents occur.

Additional examples of programming the R word are illustrated and discussed in Chapter 9. It should be noted that each machine tool and control manufacturer has its own programming specifications and requirements for its equipment. The correct programming manual should always be consulted before attempting to program a specific machine tool and control system.

RANDOM AND SEQUENTIAL TOOLING

Both random and sequential tooling are used on a variety of modern machine tools. *Random* refers to a lack of a specific pattern of tool selection.

Sequential means that tools are accessed in a particular order of succession. Sequential tool selection requires the tools to be loaded in the exact order they will be used in the program. When the program begins, the tools are selected and used one after the other, maintaining the established sequence. The correct sequential loading of the tools is of primary importance to the operator for the successful execution of the part program. If, for some reason, the tools are placed out of order, the N/C machine will not know the difference; it will just change to the next tool in sequence and place it in the spindle. Consequently, it may try to drill with a tap. This may cause significant injury to the operator, plus damaged tools and equipment. In addition, once a tool takes its turn in sequence and is returned to the tool *storage* drum, it cannot be used again. This would mean breaking the sequence to reuse a particular tool. Normally, if a tool of the same diameter is required more than once in a program, two tools of the same diameter are programmed in their proper sequence. Sequential tooling has its merits, but extreme care must be taken to maintain the proper sequence of tools used.

Random tooling is more widely used in industry because of the versatility it provides over sequential selection. When a tool change is called for, the tool changer arm removes the previous tool, puts the next tool in the spindle, and replaces the previous tool in the tool magazine in the specific pocket assigned to that tool assembly number. The *CNC,* in most cases, will remember the location of the tool by the tool assembly number initially input to the control and assigned a pocket location. The important feature of random tooling is that any tool can be accessed by the MCU and loaded into the machine spindle at any time. The MCU does not care about the order of the tools or if they have been used previously.

The programmer must be especially careful to ensure that the workpiece is moved far enough away from the spindle before a tool change is executed for either random or sequential tooling. Such precautions help prevent collisions resulting from a tool being loaded into the spindle. The programmer will often direct the machine tool to its full retracted position prior to any tool changes.

ADAPTIVE CONTROL

When programming N/C equipment, the programmer usually determines the feeds and speeds based upon the tool type and diameter, material type, setup rigidity, etc. Usually, optimum feeds and speeds are approached to make the N/C machine and cutting tool as productive as possible. This does not occur often due to excessive material hardness and dull cutting tools. Tool breakage may occur until the feeds and speeds are cut back to accommodate the particular machining circumstances.

Adaptive control, or torque controlled machining, was developed to speed up or slow down a cutting tool while the tool is engaged in the actual cutting operation. The function of adaptive control is to sense machining conditions, figure 7-11, and adjust the feeds and speeds accordingly. Sensing

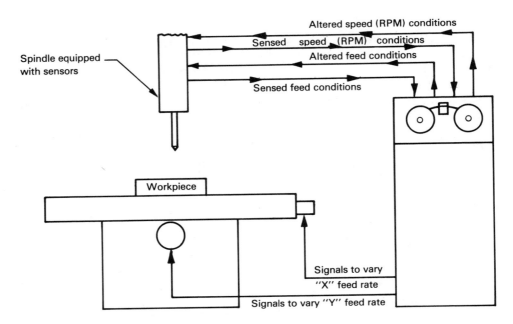

FIGURE 7-11
Diagram illustrating sensing and feedback adaptive control signals

devices are built into the machine spindle to sense torque, heat, and vibration. These sensing devices provide feedback signals to the MCU, which contains the preprogrammed safe limits. If the preprogrammed safe limits are exceeded, the MCU alters or adjusts the feeds and speeds.

Programming requirements are basically the same with adaptive control. However, it may be necessary to insert specific codes in order to turn the function on or off. The types of adaptive control and how they are used will vary, of course, among machine and control manufacturers.

The use of adaptive control is becoming increasingly popular as more companies try to optimize machine spindle time and reduce tooling requirements. Adaptive control provides automatic optimization of N/C machining operations to part manufacturing facilities.

CUTTER DIAMETER COMPENSATION

Most builders of numerical control units offer some type of *cutter diameter compensation* (CDC). This feature provides the capability to use a cutter of a different diameter than that originally intended when the part was programmed. The operator may use either an oversize or undersize cutter and still maintain the programmed part geometry.

The difference in cutter diameters, between the one programmed and the one used, can range from −1.0000 inch to +1.0000 inch. This compensation value can be input to the control in increments of 0.0001 by using the keyboard input panel. A positive (+) value indicates an oversize cutter; a negative (−) value indicates an undersize cutter.

Using this feature, a CDC value can be entered into the control for each tool number programmed. Inputting a CDC value for one tool does not affect CDC values for other tools. The CDC value becomes active when the appropriate tool is loaded into the spindle. If the value is changed after the tool is loaded, it becomes active on the next span prepared by the control system. The CDC value usually is not effective unless the appropriate codes are programmed on the tape. CDC can also be programmed with both linear and circular interpolation, but usually only in the XY plane.

Cutter diameter compensation enhances the capability of the machine and control system. It permits accurate cuts to be made with undersize or oversize cutters. Methods of input vary, but the function still provides the same increased capability.

OTHER FUNCTIONS

Many machine and control options exist which provide increased technical capabilities for both programming and operation of N/C equipment. Although it is impossible to list and discuss all options, the following terms introduce additional tooling functions on a variety of modern equipment.

TOOL TRIM

The tool trim function permits the operator to adjust the Z axis command positions to compensate for inaccuracies which could result in variations of machining depths. Tool trim codes are usually two-digit words preceded by a letter (sometimes D). The code specifies the trim value, from a group, that is to be operative during a portion of the program. The trim value is entered into the CNC by the operator. The values normally range from ±0.0001 to ±1.0000 inch. Negative values move the tool tip closer to the work surface, and positive values move it away from the surface. Tool trims for a particular tool stay in effect until cancelled by a new tool trim code, tool change, end of program, or data *reset.*

TOOL USAGE MONITOR

This feature monitors the actual tool usage time as compared to the predicted effective tool life for the particular tool. The predicted tool life can be entered by the operator via the CNC keyboard. If the tool cycle time expires while the tool is in the spindle, an error message will be displayed on the CRT screen. The machining cycle will not be inhibited but the operator will be notified that the tool should be replaced.

TOOL SETUP IDENTIFICATION

The tool setup identification feature allows tooling to be assigned to a specific set-up. Tools may also be shared between setups and common tools left resident in the tool storage mechanism. When placing a new setup on the machine or when coordinating tooling for dual fixturing or pallet shuttle machines, the CRT displays tools that are to be added or deleted during job set-up and removal.

TOOL DATA TAPE ENTRY

In addition to keyboard entry, tooling data can be entered from a punched tape and loaded into the control via the tape reader. This tape can be prepared on an off-line perforator/printer. Once loaded into the system, an updated copy of the tool data tape can also be punched directly from the control tool data file if the system is equipped with an optional tape punch feature.

REVIEW QUESTIONS

1. What is linear interpolation? What factors must be considered when programming circular cuts with linear interpolation?
2. What is circular interpolation? What four points must be programmed in circular interpolation? Why? How should arcs greater than 90 degrees be programmed using circular interpolation?
3. Name other types of interpolation available on modern MCUs and briefly discuss what they are used for.
4. Explain how some programmed Z depths are calculated. What important factors must be considered when programming Z depths?
5. What are preset tools, and why are they important? What consistency is achieved by maintaining a library of tool assemblies?
6. What is the significance of the R work surface? How is it calculated? What gage height distance is automatically allowed for when programming work surfaces?
7. When a specific work surface is programmed, what must be initiated to change the programmed work surface? Are programmed Z depths relative to a specific work surface? Explain in detail.
8. What is the order of processing a block of information which contains X-, Y-, R-, and Z-word information?
9. What types of problems may arise when changing from a lower to a higher work surface? How can these problems be avoided in programming? What must the programmer be aware of at all times?
10. Explain the difference between sequential and random tooling. Which is most widely used today? Why?
11. Describe adaptive control and its primary purpose. What advantages can be gained from adaptive control?
12. Why is cutter diameter compensation (CDC) needed? Briefly explain its function.
13. Name some other tooling control features that exist on modern N/C equipment, and briefly explain their functions.

CHAPTER 8

Modern N/C Turning Centers and Programming

OBJECTIVES   After studying this chapter, the student will be able to:

- Understand the importance of N/C turning centers to the metal-working industry.

- Discuss the traversing and positioning of the machine axes.

- Explain the basic tape controlled functions for an N/C lathe.

- Identify the types of operations performed on an N/C turning center.

- Describe system subroutines and their primary importance.

The lathe, one of the oldest and most productive machine tools, maintains its position today as an efficient producer of rotational parts. Modern N/C lathes, figures 8-1, 8-2, and 8-3, look nothing like their predecessors, but

FIGURE 8-1
A modern N/C turning center (Courtesy of Cincinnati Milicron Inc.)

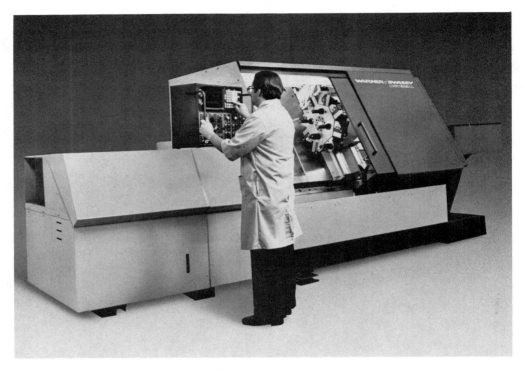

FIGURE 8-2
A modern N/C turning center (Courtesy of Turning Machine Division, The Warner & Swasey Co., subsidiary of Bendix Corporation)

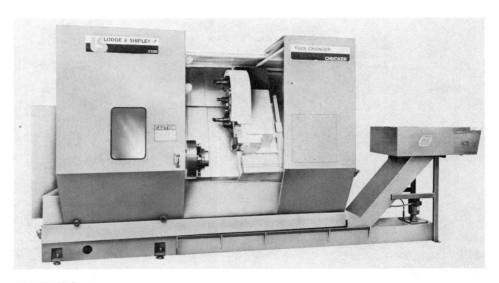

FIGURE 8-3
A modern N/C turning center (Courtesy of Lodge & Shipley Company)

they accomplish the same basic functions in a much more efficient, precise, and expedient manner. Studies indicate that the sales of N/C turning centers for the next ten years will increase substantially in the manufacturing marketplace. The increase in popularity of these mechanical and electrical engineering wonders is attributable to the competition. The ability to manufacture a product faster, better, and more economically, and to sell at a profit, has always been the motivating force of our free enterprise system. The N/C lathe is one such tool that helps to achieve that goal.

N/C LATHE AXES

A basic N/C turning center, figure 8-4, uses only two axes: Z and X. The Z axis, as mentioned before, travels parallel to the machine spindle. Therefore, a line drawn through the center of the spindle is the Z axis. For both OD (outside diameter) and ID (inside diameter) operations, a negative Z (–Z) is a movement of the saddle toward the headstock. A positive Z (+Z) is a movement of the saddle away from the headstock. The X axis travels perpendicular to the spindle centerline. A negative X (–X) moves the cross slide toward the centerline of the spindle, and a positive X (+X) moves the cross slide away from the spindle centerline. Some machine tool builders mount the cross slide on a slant bed. Other manufacturers mount their cross slides on a vertical support. Both designs allow the chips to fall free and provide very rigid support.

The N/C lathe movements are, for the most part, controlled by EIA and AIA codes, although each manufacturer uses its own specific coding format. Additional or auxiliary N/C lathe movements, such as rotating turrets, circular interpolation, and swing-up tailstocks, are also controlled through the specific tape format.

For positioning turning center axes, both absolute and incremental programming are used. It was not until recently that absolute became available. Most older turning centers were limited to incremental positioning only. Even when incremental positioning is used, the program manuscript form usually contains two extra columns (Z and X) for the programmer to keep track of the absolute dimensions from the zero point. Absolute programming, in contrast, simplifies the effort involved in manual part programming, depending on how the part is dimensioned, and helps to ensure accuracy.

Absolute and incremental programming are equally effective. For modern turning centers, the choice is up to the N/C programmer. If the dimensions on the blueprint are given incrementally, the programmer simply programs a G91 for incremental and the system immediately readies itself to accept incremental input. If workpieces are dimensioned in absolute form, the programmer uses a G90 for absolute input.

OD AND ID OPERATIONS

Regardless of the type of N/C turning center used, a variety of OD and ID operations are performed. In this discussion of OD and ID operations, we will refer to a slant bed machine with both OD and ID tooling mounted on the same turret indexing mechanism. Figure 8-4 shows a seven-position turret indexing mechanism which possesses the capacity for seven OD tools and seven ID tools. Most OD and ID tools have clearance *offsets* to avoid interference with the chuck. In addition, there is automatic compensation of the offsets when changing tools and going from an OD to an ID operation or from an ID to an OD operation.

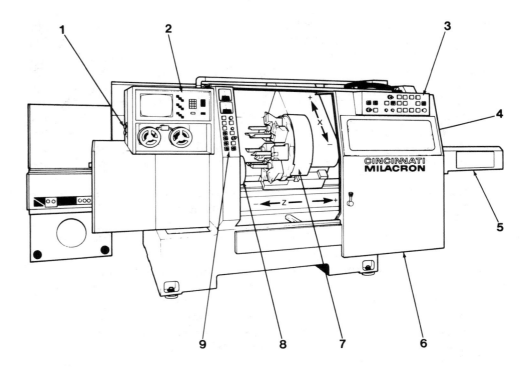

1. HI/LOW CHUCK PRESSURE
2. ACRAMATIC 900TC REMOTE CONSOLE AND TAPE READER
3. ROLLING SHIELD PANEL
4. TAILSTOCK QUILL PRESSURE (IF SUPPLIED) AND AUTO STEADY REST PRESSURE (IF SUPPLIED) ARE MOUNTED ON THE RIGHT SIDE OF BED. ROLLING SHIELD IN ILLUSTRATION HIDES THESE CONTROLS.
5. CHIP CONVEYOR
6. ROLLING SHIELD
7. TURRET
8. CHUCK
9. HEADSTOCK PANEL

FIGURE 8-4
Typical N/C turning center with axis description and major components indicated

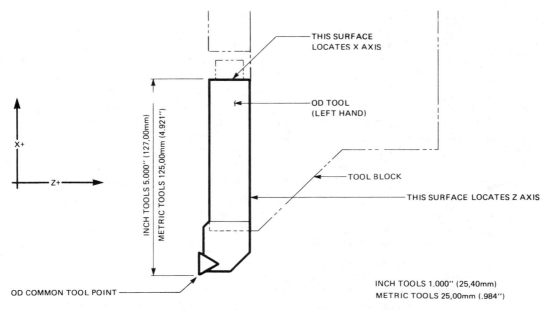

FIGURE 8-5
A qualified OD toolholder

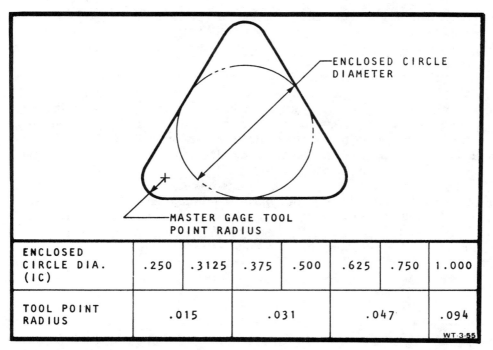

ENCLOSED CIRCLE DIA. (IC)	.250	.3125	.375	.500	.625	.750	1.000
TOOL POINT RADIUS	.015			.031		.047	.094

WT-3-55

FIGURE 8-6
Standard qualified tool point insert radii

Qualified toolholders must be used to perform OD operations. The location of the tool insert is held to close tolerances with respect to the rear and opposite sides of the toolholders. This allows holders to be changed without setting gages, figure 8-5. OD tools are "qualified" for a standard nose radius for each insert, figure 8-6. When an insert with the standard radius is used, the intersection of lines parallel to the X and Z axes and tangent to the nose radius are located in the same position for all OD tools when indexed to the machining position. This point is the common tool point. It serves as a common reference point for programming axis coordinates.

The centerline of ID tools is located on the face of the turret at the same place for inch and metric tool blocks. The centerline of ID tools is located at a fixed distance from the OD common tool point along the X axis. The distance from the ID tool centerline to the tip of the tool (intersection of tangent lines) varies with the size and construction of the particular tool. For this reason, the distance between the OD common tool point and the ID tool tip of each ID tool must be determined and compensated for when establishing program coordinates for the X axis. This is illustrated in figure 8-7.

The distance from the tip of the tool to the OD common tool point along the Z axis also varies with the size and adjustment of each tool. These distances must be determined and compensated for when determining program coordinates for the Z axis.

The distance by which each tool is offset from the OD common tool point must be established before writing the program. This must be done to determine axis coordinates and axis movements necessary to avoid interference with the tooling and workpiece. This information must be provided to the operator so the tools are set correctly.

A tool setting gage similar to the one in figure 8-8 is used to set the length of the ID tools. Scales mounted on the gage indicate the Z axis distance from the OD common tool point to the tip of the tool in both inches and millimetres. The operator mounts the tool and bushing in the gage, sets the gage block to the length specified by the programmer's instructions, and adjusts the tool to contact the gage block. The tool and bushing may then be mounted into the machine turret.

To change tools on an N/C turning center, the new turret station along with the tool change code (M06) must be programmed. The turret station and tool offsets are programmed usually with a four-digit number preceded by the letter T. The first and second digits normally designate the turret station; the third and fourth digits designate the tool offset;

Format: Txxxx

Tool offsets are then dialed in at the control panel. They are used to compensate for tool wear or for minor setup adjustments.

A variety of turning centers offer four-axis capability in order to perform simultaneous OD and ID cutting operations. Four-axis lathes with individual

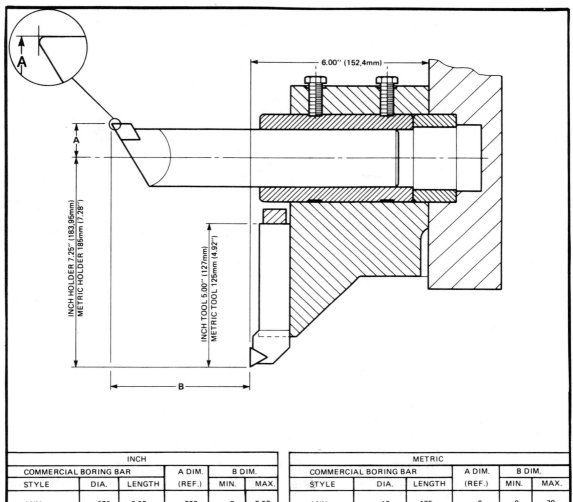

INCH					
COMMERCIAL BORING BAR			A DIM.	**B DIM.**	
STYLE	DIA.	LENGTH	(REF.)	MIN.	MAX.
ANY	.375	6.00	.250	0	2.50
ANY	.500	8.00	.312	.25	4.50
ANY	.625	8.00	.406	.25	4.50
ANY	.750	9.00	.500	1.25	5.50
ANY	1.000	9.50	.578	1.75	6.00
BORING	1.250	11.00	.765	3.25	7.00
PROFILING	1.250	11.00	.906	3.25	7.00
BORING	1.500	11.00	.890	3.25	7.00
PROFILING	1.500	11.00	1.031	3.25	7.00
BORING	1.750	12.50	1.015	4.75	8.00
PROFILING	1.750	12.50	1.156	4.75	8.00
BORING	2.000	12.50	1.281	4.75	8.00
PROFILING	2.000	12.50	1.375	4.75	8.00

METRIC					
COMMERCIAL BORING BAR			A DIM.	**B DIM.**	
STYLE	DIA.	LENGTH	(REF.)	MIN.	MAX.
ANY	10	125	6	0	70
ANY	12	140	8	0	90
ANY	16	160	10	0	110
ANY	20	230	13	35	140
ANY	25	240	17	45	150
ANY	32	280	22	85	180
ANY	40	280	27	85	180
ANY	50	320	35	125	205

FIGURE 8-7
Relationship of OD and ID tooling

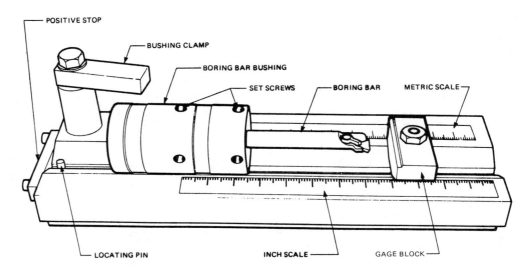

FIGURE 8-8
A typical tool setting gage

programmable slides allow considerable savings to be made because there is more than one tool in the cut at a time. This feature is illustrated in figure 8-9. The use of four-axis lathes constitutes considerable improvement in productivity over conventional turning centers (one tool in the cut at a time). However, extreme care must be exercised in programming machines of this technical complexity as the chance for errors and accidents is much higher due to two independent slide movements.

Other turning centers have two spindles with two independent slide motions for its respective OD and ID operations, figure 8-10. Machines of this nature are also capable of achieving high productivity levels with considerable savings.

It is extremely important to mention that on any N/C lathe, the turret must be positioned to a location free from interference with the chuck, workpiece, and machine elements before any tool changes are programmed. Failure to comply with this cardinal rule of N/C programming may result in bodily injury and/or machine and tool damage.

FEED RATES

The traverse rate of N/C lathe axes may be programmed in several ways. Usually these include vector rapid traverse, feed per minute, and feed per revolution. When the programmed movement requires the traversing of both axes, the axes move simultaneously along a vector path. The rate of

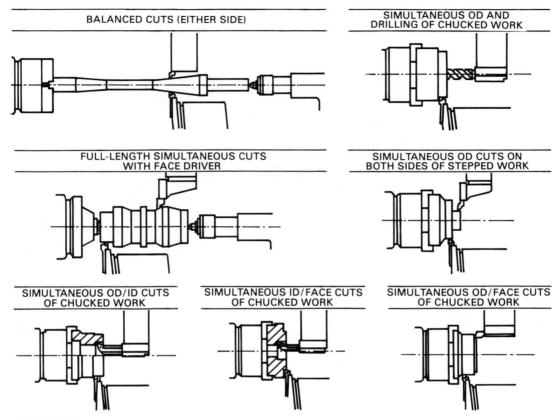

FIGURE 8-9
Four-axis N/C lathe operations showing simultaneous operations performed on a single-spindle machine

travel of each axis is set automatically by the control so the rate along the *vector* is equal to the programmed feed or rapid rate, as depicted in figure 8-11.

When a vector rapid traverse is active, the axes move simultaneously from the current position to the command position along a straight vector. Usually, the traverse rate along the vector is around 300 inches per minute. The rapid traverse rates may be modified by the operator usually by means of a *feed rate override* percent switch.

When the required preparatory function for feed per minute feed rate is programmed (G94), the feed rate is usually independent of the spindle speed. The axis feed rate for most turning centers is controlled by a four-digit number preceded by the letter F. The maximum and minimum feed rates will vary from manufacturer to manufacturer, but the ranges are somewhere around .1 to 300 inches per minute.

FIGURE 8-10
Dual-spindle N/C turning center with independent slide movements (Courtesy of Turning Machine Division, The Warner & Swasey Co., subsidiary of Bendix Corporation)

The format would be similar to the following:

Inch format — Fxxx.x
Metric format — Fxxxx.

If the programmed feed rate exceeds the allowable feed rate range, the cycle will continue but the feed rate will be set to the maximum allowable rate. If the feed rate is less than the minimum allowable rate, it will be set to the minimum allowable value.

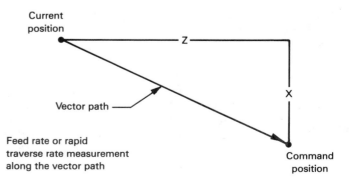

FIGURE 8-11
Vector movement

Feed rate or rapid traverse rate measurement along the vector path

Another preparatory function (G95) may be used to specify feed rate in terms of vector feed per revolution of the spindle. In this mode, the rate of axis travel varies as a function of spindle speed. The feed rate is programmed by an F word, usually in increments of .0001 inch per revolution of the spindle.

Inch format — F.xxxx
Metric format — Fx.xxx

Regarding feed per minute feed rate, if the programmed feed rate exceeds or drops below the allowable feed, the rate per revolution will be set to either the maximum or minimum allowable values.

SPINDLE SPEEDS

Spindle speeds are normally programmed on N/C turning centers with a four-digit number preceded by the letter S. The spindle speed can be programmed in either direct RPM coding or in the constant surface speed (CSS) mode.

Format: Sxxxx.   RPM
Sxxxx.   Feet per minute
Sxxxx.   Metres per minute

Maximum and minimum speeds available will vary, depending on the size and type of N/C lathe.

Other types of N/C lathes use only a two- or three-number code. The code usually refers to a table of speeds on a particular lathe. In addition, different speed ranges may be used, thereby increasing the number of available spindle speeds. Most modern CNC systems maintain a built-in system check to make sure the active spindle S word is within the allowable headstock range. A spindle speed word is also normally programmed in every block containing a headstock range change. Some N/C turning centers will display an error message, and the cycle will be stopped if the spindle speed programmed is not within the designated headstock range.

Most N/C lathes use some type of CSS feature. CSS varies the spindle automatically as a function of the X axis position to maintain the programmed value of workpiece surface speed at the point of the tool. CSS is normally input through coding of a specific G function as in direct RPM coding. The appropriate G function tells the N/C system to vary the spindle speed by speeding up, slowing down, or remaining unchanged.

FORMAT INFORMATION

Most modern controls for turning centers accept the word address, tab ignore, and variable block length tape format with either the EIA (BCD) or ASCII coding. Decimal point programming is also gaining wide acceptance. Modern CNCs will automatically sense the method used (BCD or ASCII), and will decode the tape accordingly.

The following list explains words used for a typical N/C turning center. However, not all of these words are used for every N/C system and program.

Word 1
O/N   The sequence number is composed of four digits preceded by the letter O or N (Oxxxx or Nxxxx). This word is used to indicate the block of information which is being processed by the control.

Word 2
G   The preparatory function code is a two-digit number preceded by the letter G (Gxx). These codes are used throughout the program to define the various modes of operation.

Word 3
X/Z   Axis dimensions are used to denote the position of the axes. The axes are addressed with a seven-digit number preceded by the letter X or Z. The sign denotes the direction of travel, using the incremental mode, and the position relative to program zero, using the absolute mode. (X±xxx.xxxx)

Word 4
I/K   Center point coordinates are used to define the center location when programming circular arcs. Center point coordinates are addressed with a seven-digit number preceded by a plus (+) or minus (−) sign and the letter I or K. The center point coordinates can be either absolute or incremental, depending upon the input mode. I represents the X axis, and K represents the Z axis. (I/K±xxx.xxxx)

Word 5
I/K   The axis feed rate for threading is controlled by programming a lead value. This is normally a seven-digit, unsigned number preceded by the letter I or K. Values for thread lead are not affected by the absolute or incremental input modes. The lead values are always positive, and the sign is not programmed. Programming a negative value will usually result in a program error. I represents X-axis lead, and K represents Z-axis lead. (I/Kxx.xxxxx)

Word 6
A   The rapid traverse increment is programmed with an unsigned, seven-digit number preceded by the letter A. This word is used with the automatic repeat cycle feature to define the incremental rapid approach of the tool to the work. (Axxx.xxxx)

Word 7
F   The axis feed rate is controlled by a four-digit number preceded by the letter F. Feed rates may be programmed in either distance of travel per minute or distance per revolution of the spindle, depending on the selected preparatory function. (Fxxx.x − IPM or F.xxxx − IPR)

Word 8

R The radius dimension used for CSS programming is a seven-digit number preceded by a plus (+) or minus (–) sign and the letter R. The R dimension is always an incremental value measured from the spindle centerline to the tool tip. (R±xxx.xxxx)

Word 9

V The tool retract feature is programmed with a two-digit V word. This feature programs a tool retraction along an interference-free path. The two digits of the V word represent the X and Z axes, respectively. The value of the digit that is programmed determines the direction and distance the tool will travel when the operator initiates the tool retract feature. (Vxx)

Word 10

S Spindle speeds are programmed with a four-digit number preceded by the letter S. The spindle speed can be programmed in either direct RPM coding or in the CSS mode. (Sxxxx.)

Word 11

T The turret station and offsets are programmed with a four-digit number preceded by the letter T. The first and second digits usually identify the turret station, and the third and fourth digits represent the offset. (Txxxx)

Word 12

C The C word is used to define the total number of thread starts and the thread start to be machined when machining multiple-start threads.

Word 13

D The taper trim feature compensates for taper in the workpiece. This feature is programmed using a two-digit D word. The D word designates the number of the active taper trim pair. The operator must manually input the compensation values. Both leading and trailing zeroes must be programmed. The D word uses the same format for both inch and metric values. However, it will vary with the mode selected. (Dxx)

Word 14

M The miscellaneous function codes are two-digit codes preceded by the letter M. These codes are used throughout the program to perform functions such as spindle starting and stopping, coolant control, and transmission range selection. (Mxx)

OPERATIONS PERFORMED

The absolute input mode is selected by programming a G90 word. In the absolute mode, all dimensions input into the control are referenced from a single zero point. The algebraic signs (+ and –) of absolute dimensions denote the position of the axis relative to the zero point. They do not directly specify the direction of axis travel. Some N/C units assume the G91 incremental mode when starting or when data reset operations are performed. The G90 code should be programmed at the beginning of every operation using a new tool when the program is written using the absolute mode.

The incremental input mode is selected by programming a G91 word. All dimensions input into the control are referenced from the present axis position. The input dimensions denote the distance to be moved. The algebraic sign (+ or −) in this case specifies the direction of axis travel. When an entire program is written in the incremental mode, the X and Z axes must be returned to the point where the program was started, the program start point. If this procedure is not accomplished, the axis will not be in the correct position for the start of the next workpiece. This can cause interference between the tool and workpiece or other components, resulting in tool breakage, damage to the machine, and personal injury.

LINEAR INTERPOLATION

The G01 linear interpolation preparatory function commands the slides to move the tool in a straight line from the current position to the command position. The rate of traverse is measured along the vector connecting the two points and is equal to the programmed feed rate. This mode of operation is used for turning, drilling or boring straight diameters, facing shoulders, and turning or boring chamfers and tapers.

The tool path is generated by programming the coordinates of the imaginary tool point of the tool insert radius. When turning or boring straight diameters or making facing cuts, the programmed coordinates represent tool tangent points created by constructing lines parallel to the machine axes.

When using qualified OD tools with the proper tool insert radius, the imaginary tool point and the common tool point are in the same location. When turning chamfers or tapers, the tool tangent point is at some point other than the points shown in figure 8-12. When turning a chamfer or taper, therefore, the machine axes must be offset to compensate for the new tangent point location. The amount of compensation depends on the angle of the taper and the radius of the tool insert.

Figure 8-13 illustrates the position of the imaginary tool point at the start and end of the chamfer. It also shows the path which the imaginary tool point follows during the cut. To maintain workpiece tolerances, the axes must be offset at both the start and end of the chamfer by an amount equal to A and B. When the angle of the chamfer is 45 degrees, A and B are equal.

CIRCULAR INTERPOLATION

Circular interpolation on an N/C turning center moves the tool in a circular arc along a path generated by the control system. The rate of travel

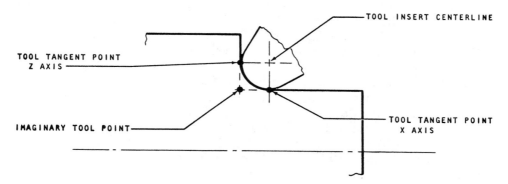

FIGURE 8-12
Tool radius tangent points — cuts parallel to a machine axis

is the same around the arc with a tangential *vector feed rate* equal to the programmed feed rate. Circular interpolation is specified by a G02 preparatory function for the clockwise direction and G03 for the counterclockwise direction. Coordinate information is also programmed to define the start point, the end point, and the center point (I = X coordinate value; K = Z coordinate value).

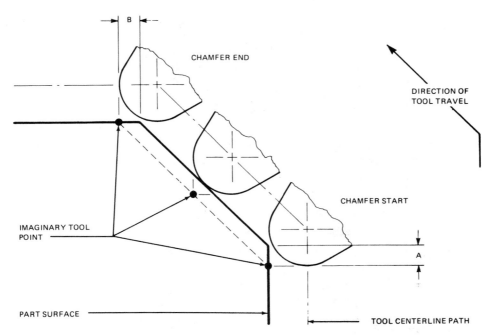

FIGURE 8-13
Imaginary tool point at start and end of a chamfer

Figure 8-14 illustrates the tool moving from the start point through a circular arc to the end point. A G02 preparatory function is used since the tool travels in a clockwise direction. At the start point, the centerline of the tool nose is placed on the arc centerline in the Z axis. The tool nose centerline is placed on the arc centerline in the X axis at the end point. While using the absolute mode, the I and K programmed center point coordinates are referenced from program zero. They are offset from the part radius center point by an amount equal to the tool nose radius. The information required to position the tool through the circular arc movement is:

N420 G02 X23188 Z20000 I23188 K29688

Figure 8-15 illustrates the tool moving from the start point through a circular arc to the end point. A G03 preparatory function is used since the tool travels in a counterclockwise direction. The tool nose centerline is placed on the arc centerline in the X axis at the start point. At the end point, the centerline of the tool nose is placed on the arc centerline of the Z axis. While using the absolute mode, the I and K programmed center point coordinates are referenced from program zero. They are offset from the part radius center point by an amount equal to the tool nose radius. The information required to position the tool to the start point is shown in block N430. Block N440 shows the information required for the circular arc movements.

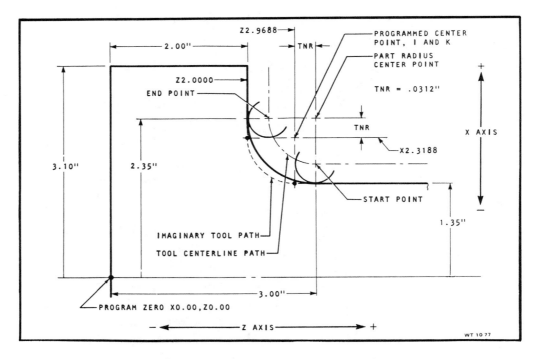

FIGURE 8-14
Inside arc — absolute mode

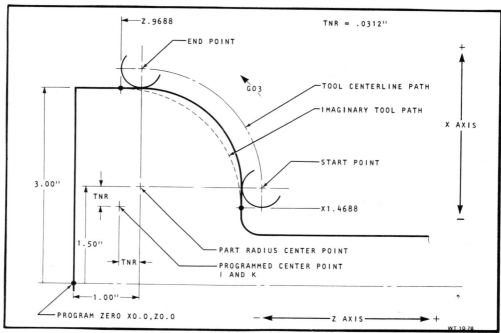

FIGURE 8-15
Outside arc — absolute mode

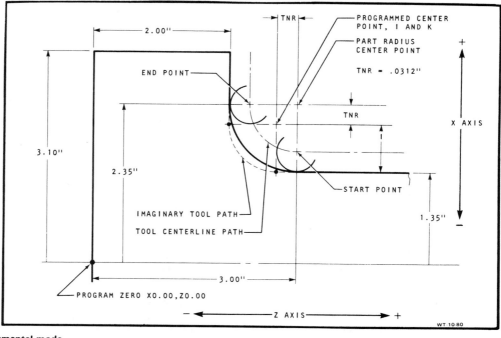

FIGURE 8-16
Inside arc — incremental mode

N430 G01 X14688 Fxxx
N440 G03 X30000 Z9688 I14688 K9688

In figure 8-16, the tool is moving through a circular arc in the incremental mode. A G02 preparatory function is used since the tool moves in a clockwise direction.

While in the incremental mode, the X axis and Z axis departure commands for 90-degree inside arcs, which start on an axis crossover point, are calculated by:

$$Xd = \text{part radius} - TNR$$
$$Zd = \text{part radius} - TNR$$

The I and K dimensions are incremental values measured from the imaginary tool point while the tool is positioned to the start point.

In this example, the I value is determined by the formula:

$$I = \text{part radius} - TNR$$
$$= 1.0000 - .0312$$
$$= .9688$$

The K value is equal to zero (KO) since the tool nose radius is on the part radius centerline in the Z axis.

The example in figure 8-17 illustrates the tool moving through an outside circular arc in incremental mode. Since the tool moves in a counterclockwise direction, a G03 preparatory function is used.

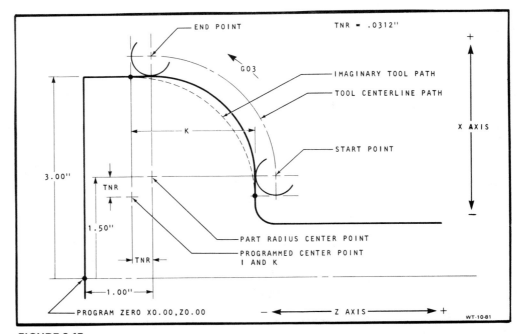

FIGURE 8-17
Outside arc — incremental mode

The X axis and Z axis incremental departure commands for 90-degree outside arcs, which start on an axis crossover point, are calculated by:

$$Xd = \text{part radius} + TNR$$
$$Zd = \text{part radius} + TNR$$

The I and K dimensions are incremental values measured from the imaginary tool point while the tool is positioned to the start point. In this example, the K value is:

$$K = \text{part radius} + TNR$$

The I value is equal to zero since the tool nose radius centerline falls on the part radius centerline in the X axis.

THREADING

Modern turning centers are capable of machining constant lead straight, tapered, and multiple-start threads. The tool is first positioned to depth and to the correct starting distance away from the workpiece. A G32 or G33 block is then programmed to cut the thread. The tool is retracted and returned for the next pass. The process is repeated, making successively deeper cuts until depth is reached. Each of these movements normally requires a separate block of information.

The slide feed rate is controlled in the constant lead threading mode by programming I and K words. I, in this case, designates lead or threads per inch of the X axis, and K designates lead or threads per inch of the Z axis.

Lead is defined as the amount the thread advances in one revolution of the spindle.

$$\text{Lead} = \frac{1}{\text{threads per inch}} \text{ or } \frac{1}{\text{threads per millimetre}}$$

Lead and the number of threads per inch are always considered to be positive. The I and K words must be programmed in every block containing a threading command.

Before any threading operation begins, the tool point must be positioned away from the workpiece before entering the thread. The minimum starting distance is:

Starting distance = (RPM X Lead X .006) + CO, where CO is compound in-feed offset, the offset generated by advancing the tool to depth or on a 29-degree angle, figure 8-18. The compound in-feed offset is calculated as:

Offset = tan 29°(Full thread depth – depth of 1st pass)

A calculation similar to the starting point offset must be performed for each threading pass when compound in-feed is used to advance the tool to depth. Figure 8-19 illustrates the compensation required for the Z axis when an .0080-inch depth of cut pass is made.

When the threads end near a shoulder, figure 8-20, space must be provided for the slide to stop. The minimum stopping distance is computed as:

Stopping distance = RPM X LEAD X .013

The following section illustrates the programming of a one-inch diameter, eight threads per inch, constant lead, single thread, and is illustrated in figure

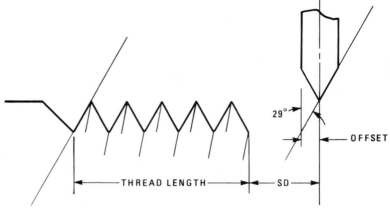

FIGURE 8-18
Starting distance — compound in-feed   SD = STARTING DISTANCE

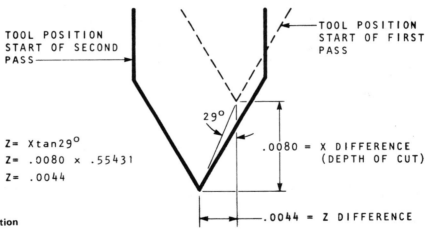

$Z = X \tan 29°$
$Z = .0080 \times .55431$
$Z = .0044$

FIGURE 8-19
Compound in-feed compensation

FIGURE 8-20
Minimum stopping distance

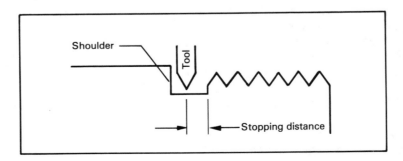

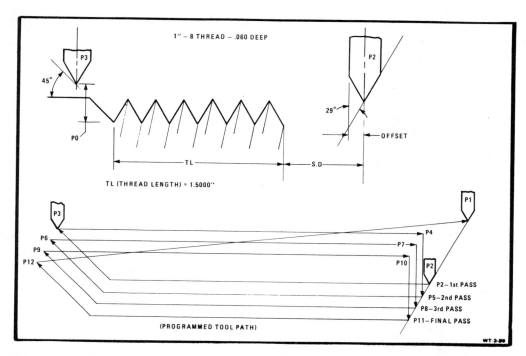

FIGURE 8-21
Example of constant lead threading showing passes required

8-21. The program example also uses incremental mode, compound in-feed of 29 degrees, and a 45-degree pullout.

Before writing the program, the following calculations are made:

Lead

$$\text{Lead} = \frac{1}{\text{threads per inch}}$$
$$= \frac{1}{8}$$
$$= .12500 \text{ (K12500)}$$

Compound Offset
CO = tan 29°(Full-thread depth − 1st pass depth
    = tan 29°(.0600 − .0300)
    = .0166

Starting Distance (with 500 RPM spindle speed)
SD = (RPM × lead × .006) + CO
    = (500 × .12500 × .006) + .0166
    = .375 + .0166

Pullout Amount – X and Z

In this example, an amount equal to the lead is used for both the X and Z axis.

X pullout = .1250
Z pullout = .1250

Z Departure Distance

$$Zd = TL + PO + (SD - CO)$$

where   TL = thread length
PO = pullout distance (Z axis)
SD = starting distance
CO = compound offset

$$Zd = 1.5000 + .1250 + (.3916 - .0166)$$
$$= 2.0000$$

X Departure = X

$$Xd = X \text{ pullout}$$
$$= .1250$$

This sample program is presented to illustrate only the tool path movements. Spindle speeds, turret indexes, and other machine-related information have been omitted from the program.

N480	G90				
N490	G00	X10000	Z50000		P1
N500	G91				
N510	G00	X-5300	Z-2938		P2
N520	G33	X1250	Z-20000	K12500	P3
N530	G00	Z19917			P4
N540	X-1400				P5
N550	G33	X1250	Z-20000	K12500	P6
N560	G00	Z19945			P7
N570	X-1350				P8
N580	G33	X1250	Z-20000	K12500	P9
N590	G00	Z19972			P10
N600	X-1300				P11
N610	G33	X1250	Z-20000	K12500	P12
N620	G00				
N630	G90				
N640	X10000	Z50000			P1

Sample Program Description:

Block N480:

The G90 is used to define the starting point P1 using the absolute mode.

Block N490:

The G00 is used to rapid the tool to P1.

Block N500:

The G91 selects the incremental input mode.

Block N510:

The tool rapids to depth for the first pass. The first cut depth is .030 inch.

Block N520:

The G33 code selects the threading mode. The X1250 command will cause a 45-degree pullout at the end of the thread. The Z command includes an additional .1250 inch for the pullout. The K word defines the lead.

Block N530:

The G00 is programmed to rapid the Z axis to P4. The Z command includes the compensation value for the 29-degree compound in-feed. This is calculated by multiplying the tan 29° by the new cut depth.

$$Zd = 2.0000 - \tan 29° \times .0150$$
$$= 2.0000 - .0083$$
$$= 1.9917$$

Block N540:

The X command rapids the tool to P5. This dimension is obtained by adding the amount of retraction to the depth of the cut for the second pass. The sign is negative since the movement is toward the centerline of the spindle.

$$Xd = .1250 + .0105 + .1400$$

Block N550:

The G33 selects the threading mode for the second pass to P6.

Block N560:

The G00 is used to rapid the Z axis to P7.

Block N570:

The X command rapids the tool to P8. The depth of cut for this pass is .010 inch.

Block N580:

The G33 is used to make the third threading pass to P9.

Block N590:

The G00 is used to rapid the Z axis to P10.

Block N600:

The tool rapids to P11. The depth of cut for this pass is .005 inch.

Block N610:

The G33 is used to make the final threading pass to P12.

Block N620:

The G00 is programmed to cancel the G33 and select rapid traverse.

Block N630:

The G90 selects the absolute input mode.

Block N640:

The X and Z commands return the tool to P1.

Some N/C turning centers provide a finish threading feature which allows for programming rough and finish threading passes, using different spindle speeds without introducing thread form errors. The threads may be rough machined at a low spindle speed and finished at a high speed. This is accomplished by compensating for changes in following error.

When a slide moves, the actual motion lags behind the command signal. This lag is characteristic of all servo systems, and is called following error. Following error is a function of feed rate. The feed rate when threading is determined by spindle speed (assuming a given lead). Changes in spindle speed will result in slight shifts of the tool position due to following error changes. This in turn produces an error in thread form.

When the finish threading feature is used, the control measures the following error of the initial pass. When the finish pass is made, the control automatically compensates slide position to hold the following error the same as the original pass when making the final passes. The thread form error is thereby eliminated.

SYSTEM SUBROUTINES

A subroutine is a set of commands or instructions that are identified and stored in the CNC system. When called upon, these instructions are put into action. The process of activating a subroutine, sometimes referred to as a program within a program, is completed by calling for this set of blocks or instructions.

Most CNC turning centers contain stored parametric variable subroutines as an optional feature which programs frequently used data, stores that data in memory, and then calls the data into action by the part program.

Once stored, the subroutine is viewed by the control in the same manner as a part program. Usually a maximum number of subroutines, part programs, or any combination of the two may be stored depending upon control specifications. The number of blocks which can be programmed in a subroutine is limited only by the total amount of storage capacity of the manufacturer's control.

Subroutines are made active by a call statement in the part program. The subroutine can be repeated many times with a single call statement and can be called any number of times by the part program.

A parametric subroutine may be programmed with variable commands so that it may be used for a variety of workpiece configurations. Ten or more variable commands can usually be used in a subroutine program. The variable commands are assigned values by the call statement of the part program.

Systems containing parametric subroutines are usually stored separate from, but in the same area as, a part program. Subroutines may be permanently or temporarily stored in memory. A permanently stored subroutine is loaded into memory by means of a multiple-program store feature. A temporarily stored subroutine is normally loaded from the beginning of the part program.

A typical example of a permanently stored subroutine is seen in figure 8-22. The subroutine defines the tool movements required to drill the part.

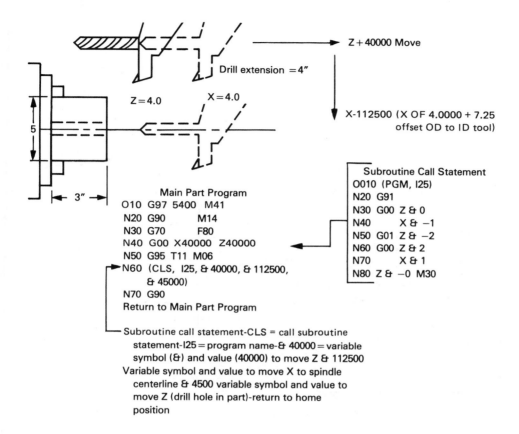

FIGURE 8-22
Example of a permanently stored parametric subroutine to offset in Z, move into position in X, drill hole in part, and return to home position (X = 4.0000 and Z = 4.0000). Subroutine may be called as many times as needed.

The subroutine call statement is programmed to repeat the subroutine as many times as called by the main program.

EXAMPLE PROGRAMS

The following programs illustrate the machining operations for typical sample parts shown in figures 8-23 and 8-24. The first example is a part illustrating a rough face, rough turn, and finish profile operation. The machining necessary to produce the second part is performed in eight operations and is illustrated in figures 8-25, 8-26, and 8-27. The second program shown is representative of basic turning, boring, and threading operations performed on an N/C turning center. It does not illustrate all phases of machining operation, programming techniques, or optional equipment.

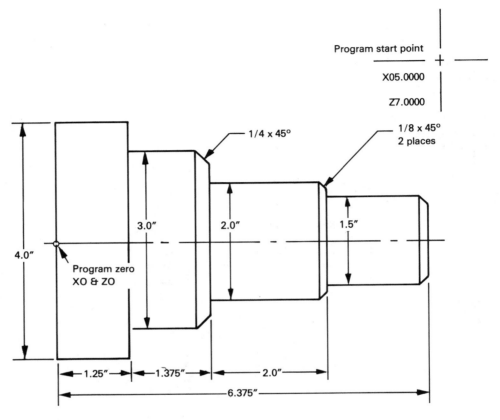

FIGURE 8-23
Example part for rough face, rough turn, and finish profile (Courtesy of Cincinnati Milacron Inc.)

BAR STK. 4.0" x 6.5
T01 (ROUGH FACE AND ROUGH TURN)
T04 FINISH PROFILE (.0468 TNR)

O10	G97 S300	M41	300 rpm — low-gear range
N20	G90	M13	Absolute positioning, spindle on w/coolant
N30	G70		Inch programming
N40	G00 X50000	Z70000	Rapid to program start pt.
N50	G95 T0101	M06	Inches per rev. — index to tool 1 + assn. offset #1
N60	G96 S400	R50000	Constant surface speed (400 sfm)
N70	G00 X22000	Z63850	Rapid to position to rough face

Example part program for figure 8-23

N80	G01	X-312	F150		Face end of stock (leave .010" for finish)
N90			Z65000		Feed away from end of stock
N100	G00	X17500			Rapid to pos. for rough turn (.25" depth) leave .010 on shldr. for finish
N110	G01		Z13500		Make 1st rough turn to 1.25 shldr.
N120		X18500	Z14500		Feed + .100 in X + Z (clearance away from part)
N130	G00		Z65750		Rapid out in Z axis .200 from end of part
N140		X16000			Rapid down to next depth of cut
N150	G01		Z13500		Make second rough turn to 1.25 shldr.
N160		X16100	Z14500		Feed + .100 in X + Z (clearance)
N170	G00		Z65750		Rapid out in Z .200 from end of part
N180		X12500			Rapid down to next depth
N190	G01		Z27250		Make rough turn to 2.625 shldr.
N200		X16100	Z22750		Rough 1/4 x 45° chamfer
N210	G00		Z65750		Rapid out .200 from end of part
N220		X11000			Rapid down to next depth
N230	G01		Z27250		Make rough turn to 2.625 shldr.
N240		X12000	Z28250		Feed + .100 in X + Z (clearance)
N250	G00		Z65750		Rapid out .200 from end of part
N260		X7250			Rapid to next depth
N270	G01		Z64750		Feed in to .100 from end of part
N280		X8500	Z63500		Rough 1/8 chamfer on end of part
N290			Z47250		Feed back to 4.625 shldr.
N295		X8850			Feed up 4.625 shldr.
N300		X11000	Z44000		Rough 1/8 x 45 chamfer on 4.625 shldr.
N310		X12000			Feed up .100 to clear part
N320	G00	X50000	Z70000		Rapid turret back to prog. st. pt.
N330	G97	S1000	M42		Direct rpm (1000) — high-gear range
N340	G90				Absolute positioning
N350	G70				Inch programming
N360	G00	X50000	Z70000		Restate current pos. for tool index
N370	G95	T0404	M06		Inches per rev. — index to tool #4 + assn. offset #4
N380	G96	S1000	R50000		Constant surface speed (1000 sfm)
N390	G00	X-470	M13	Z65750	Rapid to ℄ and .200 from end of part
N400	G01	F80	Z63750		Feed in to end of part to begin finish profile
N410		X5976			Feed from ℄ to start of chamfer on end of part
N420		X7500	Z62226		Cuts chamfer on end of part (1/8 x 45°)
N430			Z46250		Feed back in Z to 4.625 shldr.
N440		X8476			Feeds up 4.625 shldr. to start of 2nd 1/8 x 45°C
N450		X10000	Z44726		Cuts 1/8 x 45°C on 4.625 shldr.
N460			Z26250		Feeds across 2.0" dia. to 2.625 shldr.
N470		X12226			Feeds up 2.625 shldr. to start of 1/4 C
N480		X15000	Z23476		Cuts 1/4 x 45°C up 3.0" dia.
N490			Z12500		Feeds across 3.0" dia. to 1.25 shldr.
N500		X21000			Feeds up 1.25" shldr. + clears str. dia. by .100"
N510	G00	X50000	Z70000		Rapid back to st. pt.
N520		T0100	M06		Cancel out active assignable offset
N530		M30			Ends program (M30 shuts off spindle and coolant — also rewinds program)

Example part program for figure 8-23 (Continued)

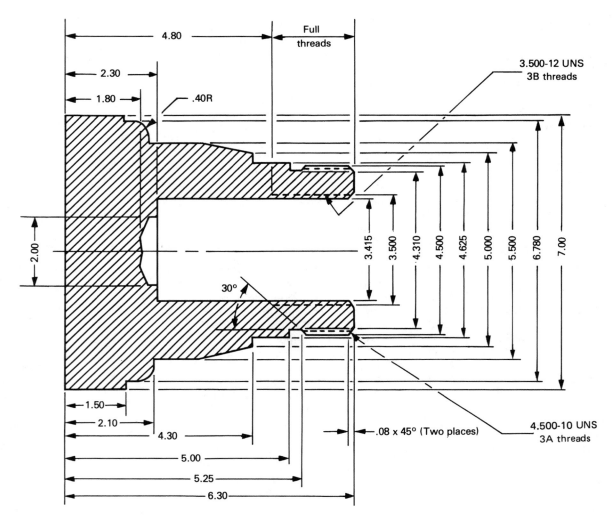

FIGURE 8-24
Sample part engineering drawing

First Operation	O10	G90			
	N20	G97	S100	M42	
	N30	G70	M03		
	N40	G00	X50000	Z85000	T0100	M06
	N50	G95			
	N60	G92	S2500		
	N70	G96	R50000	S600	
	N80	G00	X37000	Z63500	M08
	N90	G01	X-940	F150	
	N100	Z65500	F600		
	N110	X37000			
	N120	Z63000			
	N130	X-940	F150		
	N140	G00	Z65000		
	N150	X45000			
Second Operation	O160	G90			
	N170	G97	S351	M41	
	N180	G70	M13		
	N190	G00	X45000	Z65000	T0200	M06
	N200	G95			
	N210	G92	S2500		
	N220	G96	R45000	S600	
	N230	G00	X32600		
	N240	G01	Z19690	F150	
	N250	G03	X34100	Z16062	I28962	K16062
	N260	G01	X37000		
	N270	G00	Z65000		
	N280	X30100			
	N290	G01	Z21200		
	N300	X32100			
	N310	G00	Z65000		
	N320	X27600			
	N330	G01	Z21100		
	N340	X28962			
	N350	G03	X34000	Z16062	I28962	K16062
	N360	G01	Z15100		
	N370	X36000			
	N380	G00	X29600	Z65000	
	N390	X25100			
	N400	G01	Z42199		
	N410	X27600	Z32986		
	N420	G00	Z65000		
	N430	X22600			
	N440	G01	Z55989		
	N450	X21650	Z51659		
	N460	Z50100			
	N470	X23225			
	N480	Z43100			

Sample part program for figure 8-24

	N490	X26000				
	N500	G00	Z130000			
Third Operation	O510	G90				
	N520	G97	S600	M41		
	N530	G70	M14			
	N540	G00	X26000	Z130000	T1100	M06
	N550	G95				
	N560	G00	X-72500			
	N570	G01	Z83000	F150		
	N580	G00	Z148000			
Fourth Operation	O590	G90	M05			
	N600	G97	S1000	M42		
	N610	G70	M13			
	N620	G00	X-72500	Z148000	T1200	M06
	N630	G95				
	N640	G92	S2500			
	N650	G96	R10310	S600		
	N660	G00	X-71560	Z115000		
	N670	G01	Z73100	F150		
	N680	X-73750				
	N690	G00	Z115000			
	N700	X-70310				
	N710	G01	Z73100			
	N720	X-72500				
	N730	G00	Z115000			
	N740	X-69060				
	N750	G01	Z73100			
	N760	X-71250				
	N770	G00	Z115000			
	N780	X-67810				
	N790	G01	Z73100			
	N800	X-70000				
	N810	G00	Z115000			
	N820	X-66560				
	N830	G01	Z73100			
	N840	X-68750				
	N850	G00	Z115000			
	N860	X-65835				
	N870	G01	Z73100			
	N880	X-68023				
	N890	G00	Z148000			
Fifth Operation	O900	G90				
	N910	G97	S846	M42		
	N920	G70	M13			
	N930	G00	X-68023	Z148000	T1300	M06
	N940	G95				
	N950	G92	S2500			
	N960	G96	R14787	S800		

Sample part program for figure 8-24 (Continued)

```
                        N970    G00  X-64752  Z115000
                        N980    G01  X-65735  Z112017    F100
                        N990    Z73000
                        N1000   X-73210
                        N1010   G00  Z148000
Sixth Operation         O1020   G90  M05
                        N1030   G97  S400    M41
                        N1040   G70  M14
                        N1050   G00  X-73210  Z148000    T1400   M06
                        N1060   X-67805  Z103000
                        N1070   G91
                        N1080   G33  X-1500   Z-18668  K8333
                        N1090   G00  Z18613
                        N1100   X1600
                        N1110   G33  X-1500   Z-18668  K8333
                        N1120   G00  Z18629
                        N1130   X1570
                        N1140   G33  X-1500   Z-18668  K8333
                        N1150   G00  Z18640
                        N1160   X1550
                        N1170   G33  X-1500   Z-18668  K8333
                        N1180   G00  Z18646
                        N1190   X1550
                        N1200   G33  X-1500   Z-18668  K8333
                        N1210   G00  Z18654
                        N1220   X1525
                        N1230   G33  X-1500   Z-18668  K8333
                        N1240   G00  Z18670
                        N1250   X1520
                        N1260   G33  X-1500   Z-18668  K8333
                        N1270   G00
                        N1280   G90
                        N1290   Z105000
                        N1300   X45000
Seventh Operation       O1310   G90
                        N1320   G97  S780    M41
                        N1330   G70  M13
                        N1340   G00  X45000   Z105000    T0300   M06
                        N1350   G95
                        N1360   G92  S2500
                        N1370   G96  R45000   S800
                        N1380   G00  X19517   Z65000
                        N1390   G01  X22500   Z62017    F100
                        N1400   Z54373
                        N1410   X21550   Z52728
                        N1420   Z50000
                        N1430   X23125
                        N1440   Z43000
```

Sample part program for figure 8-24 (Continued)

```
                         N1450   X24927
                         N1460   X27500   Z33399
                         N1470   Z21000
                         N1480   X29588
                         N1490   G03  X33900   Z16688   I29588  K16688
                         N1500   G01  Z15000
                         N1510   X37000
                         N1520   G00  X45000   Z65000
Eighth Operation         O1530   G90
                         N1540   G97  S300   M41
                         N1550   G70  M14
                         N1560   G00  X45000   Z65000   T0400   M06
                         N1570   X22300   Z66029
                         N1580   G33  Z50300   K10000
                         N1590   G00  X24500
                         N1600   Z65946
                         N1610   X22150
                         N1620   G33  Z50300   K10000
                         N1630   G00  X24500
                         N1640   Z65879
                         N1650   X22030
                         N1660   G33  Z50300   K10000
                         N1670   G00  X24500
                         N1680   Z65835
                         N1690   X21950
                         N1700   G33  Z50300   K10000
                         N1710   G00  X24500
                         N1720   Z65807
                         N1730   X21900
                         N1740   G33  Z50300   K10000
                         N1750   G00  X24500
                         N1760   Z65791
                         N1770   X21870
                         N1780   G33  Z50300   K10000
                         N1790   G00  X24500
                         N1800   Z65780
                         N1810   X21850
                         N1820   G33  Z50300   K10000
                         N1830   G00  X24500
                         N1840   X50000   Z85000
                         N1850   M30
```

Sample part program for figure 8-24
(Continued)

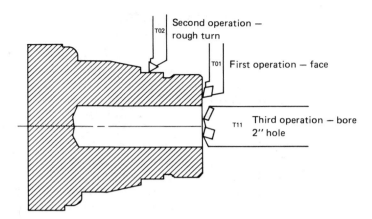

FIGURE 8-25
First, second, and third operations

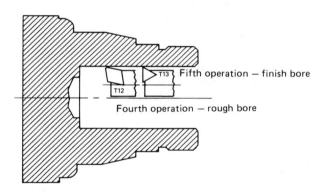

FIGURE 8-26
Fourth and fifth operations

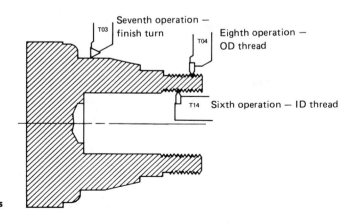

FIGURE 8-27
Sixth, seventh, and eighth operations

REVIEW QUESTIONS

1. Explain the Z-axis and X-axis relationships on an N/C lathe, including positive and negative moves in each direction.
2. Explain the differences between absolute and incremental programming on an N/C turning center. What is a major factor in determining whether to program in an absolute or an incremental mode?
3. How are OD and ID tools accurately located in the tool turret?
4. Discuss the importance of tool offsets. Why are they used?
5. Why is it necessary to know the location of the tip of the tool prior to programming a tool change?
6. Explain the difference between the three types of axis feed rates, and discuss what takes place when a programmed feed rate exceeds an allowable feed rate range.
7. What is meant by constant surface speed (CSS)? Briefly explain how it functions.
8. What turning center operations are performed with linear interpolation moves? How is the tool path information generated?
9. Why is circular interpolation used on N/C turning centers? What are the basic differences between programming an incremental or absolute circular interpolation move?
10. Why is it important to program Z- and X-axes leads on a threading operation? Are thread lead values affected by absolute or incremental input modes?
11. Why should a space be provided in the part when threading is ended at a shoulder?
12. What is finish threading? How is it accomplished on an N/C turning center?
13. Briefly explain a system subroutine. Are any limitations placed on the number of blocks that are programmed in a subroutine? How?
14. What is a parametric subroutine? What is the difference between a temporary and a permanent stored subroutine? How are each accessed by the N/C part program?

CHAPTER 9

N/C Machining Centers and Programming

OBJECTIVES After studying this chapter, the student will be able to:

- Understand the importance and versatility of N/C machining centers.
- Explain automatic tool changing and tool storage capabilities.
- Discuss the advantages, capabilities, and versatility of a rotary index table.
- Understand preset tool lengths/compensation and their use.
- Describe the types of operations performed on an N/C machining center.

TYPES OF TOOL CHANGERS

Numerically controlled tool changers are still considered "state of the art" in N/C machining. This is because they change their own tools without operator intervention. Basically, there are two types of tool changers: vertical and horizontal.

Vertical tool changers, figure 9-1a and 9-1b, are unique in their own way, but do not possess the overall advantages of horizontal machining centers. Horizontal tool changers, commonly referred to as *machining centers,* are shown in figure 9-2 and 9-3. The machining center has made the greatest impact in N/C machine tool design and concept. It will perform a number of different operations such as milling, drilling, boring, spotting, counterboring, and tapping in a single setup of workpiece. In addition, the machine will change its own tools. It is this aspect that has placed the horizontal machining center in the limelight. Undoubtedly, this is where much emphasis will be placed in the years ahead.

Machining centers are capable of machining on all sides of a workpiece in one setup. Some machining centers have been developed which combine typical tool-spindle operations with workpiece-turning operations. Theoretically, there is no limit as to how much can be combined in a single machine. It is merely a question of how much engineering and development effort can be applied.

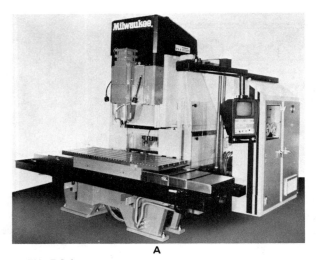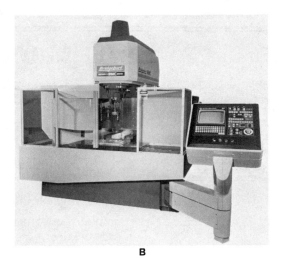

A   B

FIGURE 9-1
Vertical tool changers (A, Courtesy of Kearney & Trecker Corporation; B, Courtesy of Bridgeport Machines)

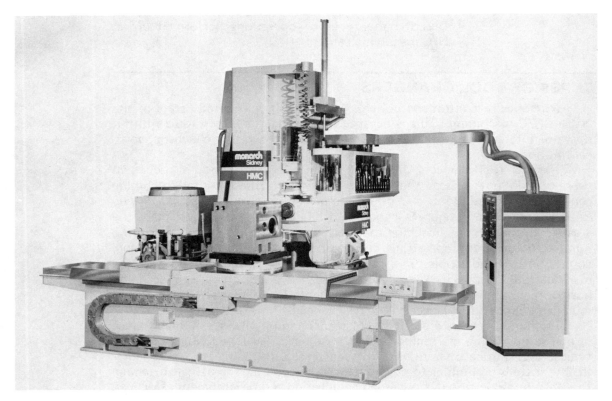

FIGURE 9-2
A typical N/C machining center (Courtesy of Monarch Machine Tool Company)

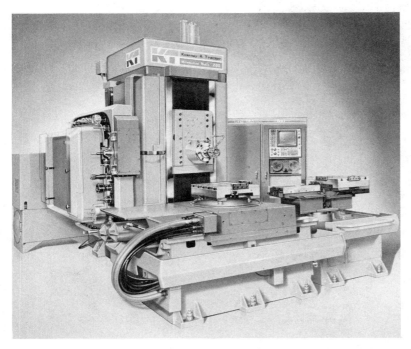

FIGURE 9-3
A typical horizontal machining center (Courtesy of Kearney & Trecker Corporation)

Typical motions of horizontal machining centers include the X, Y, Z, and B motions. However, more complex machining centers are capable of many other axis movements. These movements may include tilt and swivel of the spindle head and column and rotation of the workpiece against a fixed tool, such as a lathe turning operation.

N/C machining centers represent a real frontier for future development. The unlimited potential of multiaxis capabilities will provide new and better ways to locate and machine various types of workpieces in the future.

TOOL STORAGE CAPACITIES

As mentioned, one advantage of tool changers is that they automatically change their own cutting tools. This capability is applied to both vertical and horizontal applications; the tool magazine may have a vertical or horizontal axis. The tool magazine may then rotate so that the center of the tool is automatically aligned with the spindle. In most cases, the tool magazine is to one side or above the spindle. Figure 9-4 illustrates a vertical tool magazine and changing mechanism. Figure 9-5 shows a horizontal tool magazine and changing mechanism. A few machines have the tool magazine at an angle to the spindle axis.

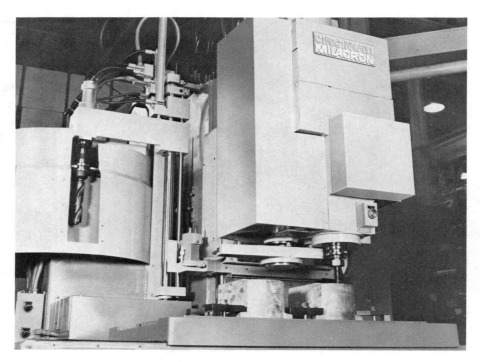

FIGURE 9-4
A vertical tool magazine and changing mechanism (Courtesy of Cincinnati Milacron Inc.)

The tools may be selected sequentially (in order of use or succession), or they may be selected at random. Numbers may then be assigned to the position or pocket on the tool. This means that the programmer must, in some cases, decide which pockets or tool positions to use. The programmer also must supply the setup person with a tool list identifying which cutting tool is to be used in each position. Certain tools may be assigned permanent positions in the tool magazine, depending on the frequency of use. This leaves the remaining pockets available for special tools.

The actual tool magazine for most machining centers and tool changers has holes around the perimeter of a circle, or on a chain, spaced at a specific distance. The maximum diameter of any cutter held in magazine storage cannot be greater than the hole spacing if all pockets are to be filled.

Tool-changing mechanisms vary greatly. Some have a combined operation of four machine elements to change tools.

- *Tool drum* contains thirty or more coded drum stations. It rotates in the direction commanded by the control.

- *Intermediate transfer arm* removes tools from the tool drum and places them in the interchange arm and then returns them to the tool drum.

- *Interchange station arm* receives tools from the intermediate transfer arm and swings them forward into the proper position for a tool interchange

FIGURE 9-5
A horizontal tool magazine and changing mechanism (Courtesy of Cincinnati Milacron Inc.)

with the tool changer arm. It then swings them back into position for removal by the immediate transfer arm.

- *Tool changer arm* simultaneously removes a tool from the interchange station arm and one from the spindle. It then interchanges these tools and inserts them into the spindle nose and the interchange station arm.

One of the most important considerations of tool changers, regardless of the type, is whether the tool being removed from the spindle or the new tool going into the spindle will clear the workpiece and any other obstructions such as clamps and pushers. When programming with longer tools, the programmer must be extremely careful that the workpiece is moved far enough away from the spindle so that no collision will occur when the longer tool is inserted into the spindle. Often a retract to the extreme rear position in Z or an offsetting move in X or Y is required to avoid collisions.

TOOL LENGTH STORAGE/COMPENSATION

Tool length storage and *tool length compensation* allows the control to store information relative to a given tool length. This stored length is then applied to the Z axis position when the tool is loaded into the spindle by the tool change code (M06). This helps the programmer to program without knowing exact tool lengths and helps eliminate errors in calculating both the Z slide and tool length.

Practice has shown that the programmer must supply the operator with certain agreed-upon information to set up a job. With tool length storage/ compensation, the following methods have worked very well.

Method 1
  a) The programmer makes a tool assembly drawing for each tool used, showing all the components of a tool assembly, figure 9-6. The tool set length is calculated and rounded off to the nearest one-eighth inch. When the tool is set up, the actual set length must be within a tolerance of the length dictated by the programmer .
  b) The position where tool length values will be established by the programmer and the distance from this point to the centerline of index is determined. This distance and the feeler gage thickness are added to define a tool tram surface value. The operator enters this value, assigned by the programmer, into the control before setting tool length values.
  c) After the operator loads the tools into the tool drum, each tool is located in the spindle and touched up to the tool tram surface. On some controls, depressing the TOOL LENGTH COMP. SET push button will initiate the calculation of a tool length and will store that information under the data for the tool.

Method 2
  a) The programmer makes a tool assembly drawing for each tool used (figure 9-6), showing all the components of the tool assembly. The tool set length is calculated and rounded off to the nearest one-eighth inch. When the tool is set up, the actual set length must be within a tolerance of the length dictated by the programmer.
  b) The operator or tool specialist gages the exact length of each tool by means of a tool preset gage and makes note of it.
  c) The operator loads tools into the tool drum and enters the length values for the tools into the control under their respective tool data information.

Tool setting is not difficult since there are several simple mechanical and optical tool-setting devices. Sometimes the tools and toolholders are set by the toolroom, and the machine operator, having the time and ability, makes up the preset tools.

Many shops maintain tool assembly drawings. These drawings show the cutter, tool number, and setting distance. They simplify tool setting and provide consistent accuracy to all tools assembled. For example, every time a 10.375 set length, .500-inch diameter drill is used, the part programmer can check the drawing and find that a 10.375 set length, .500-inch drill has a particular tool assembly number associated with it. The part programmer then calls for this tool assembly number in the part program. With this information, all part programmers can write programs calling for various tool assembly numbers and feel confident that everyone has the same dimensions.

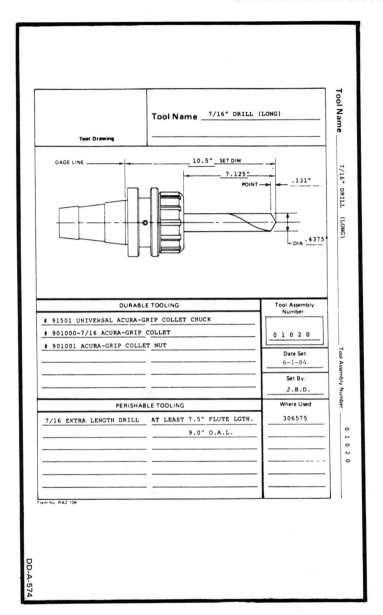

FIGURE 9-6
A typical tool assembly drawing (Courtesy of Cincinnati Milacron Inc.)

WORK TABLES

Rotary index tables are another feature of horizontal N/C machining centers that provide versatility. With proper fixturing and one clamping of the workpiece, the entire part can be machined in one setting. Rotary tables are usually designated by the *B (beta) axis* and must be aligned the same as the X, Y, and Z axes. An example of a rotary index table and all axes orientation can be seen in figure 9-7.

Some rotary index tables are equipped with a universal fixture base and right-angle plate. These options maximize the productivity of a machining center. A conventional index table can be seen in figure 9-8.

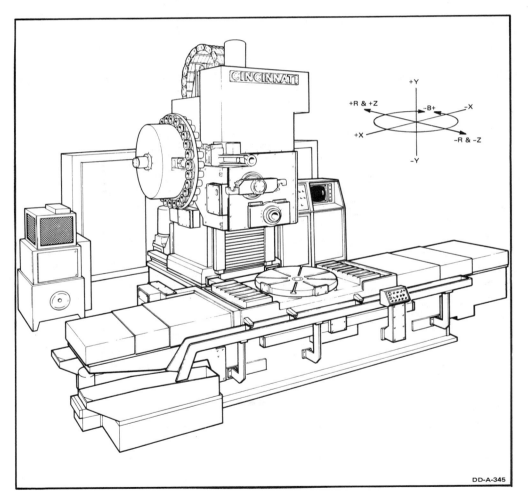

DD-A-345

FIGURE 9-7
Rotary index table and machine with axis orientation (Courtesy of Cincinnati Milacron Inc.)

FIGURE 9-8
A conventional index table with part mounted on machine table (Courtesy of Monarch Machine Tool Company)

There are two basic types of rotary tables. However, the possibility of different methods of operation and control provides many combinations. The first type of rotary table uses a positive, serrated plate to position the table mechanically. This rotary table will lift before indexing and lower into its position after indexing. The second type of rotary table uses a rotary inductosyn seal to position and provide some means of feed rate control. This type of table may even be interpolated with the other slides to provide four-axis contouring.

Most rotary tables are bidirectional and will index using the shortest path to any of 72, 360, or 720 positions. The different degrees of rotation in a rotary index table are illustrated in figure 9-9. The input is in degrees and, in most cases, all positions are in absolute positioning. Some rotary tables can index up to 360,000 positions. These tables are programmable either in absolute or incremental modes. Methods of programming will vary with each manufacturer.

As much as rotary index tables can increase a machine tool's versatility and productivity, they can

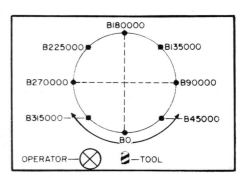

FIGURE 9-9
Rotary index table illustrating degrees of rotation

also be a real danger if certain considerations are not met: 1) The programmer should always make sure the tool is located in a safe position so the table will index without hitting the cutting tool; 2) Two blocks of information should be programmed to ensure the desired direction of table rotation when a 180-degree rotation is required, if the direction of rotation has a possible interference.

A few facts, and some tooling tips, are worth remembering and repeating. A machining center can put more different tools into the workpiece in a specified time period than any conventional machine tool. Thus, total tool usage is generally greater. With so many cutting tools being applied without guidance of bushings, the cutting tools must have symmetry. Keep in mind that with all the precision built into the machine, it is no more accurate than the cutting tools used to machine the workpiece.

FORMAT INFORMATION

There are many different types of machining centers and tape formats available. Because of this variety, it would be impossible to discuss all types. However, to illustrate machining center capacity from a conceptual point of view, a review of what is actually programmed on a typical machining center is in order.

Most modern controls for machining centers accept the word address, tab ignore, variable block length, or decimal point programming format with either the BCD or ASCII coding. Most controls will automatically sense which method is used and will decode the tape accordingly. A typical programming form, data sequence, and tape format is shown in figure 9-10. The minimum increment of input for this particular format is 0.0001 inch or 0.001 millimetre. Not all of the words illustrated in this figure are used for every system and program. Nevertheless, the following is a brief explanation of each word, the meaning of each character in the word, and their use as they appear in this typical machining center tape format.

		PART NAME						PART NO.				REVISION	
NC PROGRAMMING SHEET		MACHINE						PROG. BY	CHECKED BY		DATE	PAGE	
		SETUP INFORMATION											
O/N SEQ.	**G** PREP FUNCT.	**X** ± POSITION	**Y** ± POSITION	**Z** ± POSITION	**R** ± POSITION	**I/J/K** POSITION		**A/B/C** POSITION **P/Q** ± WORD		**F/E** FEED RATE	**S** SPINDLE SPEED	**D** WORD	**T** TOOL WORD	**M** MISC. FUNCT.
Ø----	G--	X±-------	Y±-------	Z±-------	R±--------	I±-------	J±-------	P±-----	Q±-----	F----	S----	D--	T--------	M--

FIGURE 9-10
Typical machining center tape format

P - dwel # sec.
Q - Pek draw

Word 1
O/N   Sequence number coding, introduced by O or N. It is a five-character code: one letter and four numerals (Oxxxx, or Nxxxx). It is used to identify a block of information. It is informational, rather than functional.

Word 2
G   Preparatory function coding, introduced by G. It is a three-character code: one letter and two numerals (Gxx). It is used for control of the machine. It is a command that determines the mode of operation of the system. This word is informational and functional. Therefore, all characters, except leading zeros, must be included in the code.

Word 3
X   X axis coordinate information code, introduced by X. It may contain up to nine characters: one letter, one sign, and up to seven numerals (X xxxx.xxx millimetres). Six numerals are normally used. This word is used to control the direction of table travel and position.

Word 4
Y   Y axis coordinate information coding, introduced by Y. This is identical to Word 3, only using the Y address.

Word 5
Z   Z axis coordinate information coding, introduced by Z. This is identical to Word 3, only using the Z address.

Word 6
R   Z axis coordinate information coding, introduced by R. It may contain up to nine characters: one letter, one sign, and up to seven numerals (R xxx.xxxx inches, R xxxx.xxx millimetres). It is used to control the positions of the Z slide at rapid traverse during positioning mode.

Word 7
I   The center point coordinate in circular interpolation of the X axis is introduced by I. It may contain up to nine characters in the coding: one letter, one sign, and up to seven numerals (I xxx.xxxx inches, I xxxx.xxx millimetres).

Word 8
J   The center point coordinate in circular interpolation of the Y axis is introduced by J. This is identical to Word 7, only using the J address.

Word 9
K   The center point coordinate in circular interpolation of the Z axis is introduced by K. This is also identical to Word 7, only using the K address.

Word 10
B   Beta axis coding of the *index table,* introduced by B. This is a seven-character code: one letter and six numerals (Bxxx.xxx). It determines the angular position of the index table.

Word 11

P   The X axis projection, of the cutter diameter compensation vector, introduced by P. It is a seven-character code: one letter, one sign, and five numerals (Px.xxxx). It describes the CDC unit vector value for the X axis (optional feature).

Word 12

Q   The Y axis projection, of the cutter diameter compensation vector, introduced by Q. This is identical to Word 11, only using the Q address (optional feature).

Word 13

F   Feed rate coding for X, Y and/or Z axes, introduced by F. It may contain up to five characters in the code: one letter and up to four numerals (Fxxx.x inches/minute, Fxxxx millimetres/minute). It is used for controlling the rate of longitudinal, vertical, and cross travel.

         The F word is also used in conjunction with the G04 code to dwell the slides. In this mode, the format can vary from .01 second to 99.99 seconds of dwell.

Word 14

S   Spindle speed coding for rate of rotation of the cutting tool, introduced by S. It may contain up to five characters in the code: one letter and up to four numerals (Sxxxx). This is the actual RPM desired for cutting.

Word 15

D   Tool trim coding of the tool axis is introduced by D. It may contain up to three characters in the code: one letter and up to two numerals (Dxx). This selects the stored value to be applied to the Z axis positions for a given operation (optional feature).

Word 16

T   Tool number coding introduced by T. It is a nine-character code: one letter and eight numerals (Txxxxxxxx). It determines the next tool to be used.

Word 17

M   Miscellaneous function coding, introduced by M. It is a three-character code: one letter and two numerals (Mxx). It is used for various discrete machine functions.

OPERATIONS PERFORMED

Many different types of operations are performed repeatedly by machining centers. Therefore, it is important to examine and thoroughly understand the basic operations performed on a machining center and how a typical machining center format appears.

DRILLING

The concept of drilling is an old, reliable method of metal removal, regardless of what machine is to perform the operation. However, some helpful hints are worth mentioning.

Use the shortest drill possible to accomplish the job. They are more rigid and are capable of greater accuracy. Lip height, clearance, and angle of point must be ground accurately for best results.

The following axis movements will occur when a G81 drill cycle is programmed:

1) rapid in X and Y;
2) rapid the Z axis to gage height;
3) feed in the Z axis to gage depth; and
4) rapid retract to gage height.

The use of the G81 drill cycle is illustrated in figures 9-11, 9-12, and 9-13.

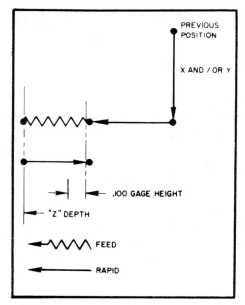

FIGURE 9-11
G81 — Fixed cycle drill schematic

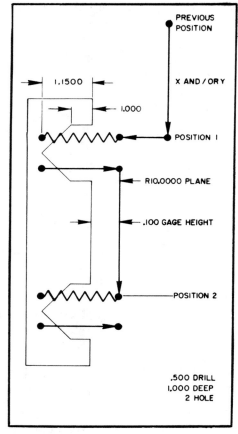

FIGURE 9-12
G81 — Fixed cycle drill — cycle representation

O/N SEQ.	G PREP. FUNCT.	X± POSITION	Y± POSITION	Z± POSITION	R± POSITION	I/J/K POSITION		A/B/C POSITION P/Q ± WORD		F/E FEED RATE	S SPINDLE SPEED	D WORD	T TOOL WORD	M MISC. FUNCT.
Ø 15	G81	X+ 40000	Y+100000	Z–11500	R+100000			B 0		F 100	S550		T 3	M03
N 16			Y 80000											

FIGURE 9-13
G-81 — Sample program

In sequence number O15, the G81 code rapid advances the X and/or Y axes simultaneously to Pos. 1 from the previous position. When Pos. 1 is reached, the Z axis will rapid to the R 10.0000 plane (gage height). At this point the Z axis will feed to the programmed depth* at the programmed rate.

After reaching depth, the Z axis will rapid retract to the R 10.0000 plane (gage height), and the next block of information (N16) will be read and acted upon.

In sequence number N16, the G81 code is used again to rapid the Y axis, with the tool at gage height, to Pos. 2. Then the Z axis will feed to the programmed depth* at the programmed rate. After reaching depth, the Z axis will rapid retract to the R 10.0000 plane (gage height).

MILLING

Milling is widely used on all types of machining centers. Some brief suggestions prior to an application of milling on a machining center are necessary.

The cutter and the workpiece should always be placed as close as possible to the spindle nose and table of the machine. Repetitive accuracy of work requires rigid locating surfaces. The workpiece should be properly supported and clamped against these locating surfaces. A stop should be placed at one end to oppose the thrust of the cutting load.

The following axes movements, as shown in figure 9-14, will occur when the G79 (basic mill cycle) is programmed. Figure 9-14 illustrates the G79 basic cycle as used for Z motion combined with X and Y motion. When G79 is programmed, the axes will:

1) feed in the X and/or Y axes (linearly interpolated);
2) feed in the Z axis to the R plane; and then
3) feed in the Z axis to the Z depth.

---

*Programmed depth for the Z axis is calculated as follows:

Z position   Depth of cut + drill point
              1.0000 + (0.3 × diameter of drill)
              1.0000 + (0.3 × 0.5000)
              1.1500

(The value of 0.3 is used for a standard 118-degree drill point.)

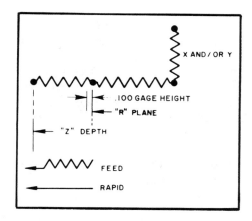

FIGURE 9-14
G79 — Basic mill cycle schematic

These three steps will occur in the same order every time a G79 cycle is programmed.

In the first two steps, if the movement is zero or has already been satisfied by a previous block of information, actual movement of the slide, or tool, for that step of the cycle will not occur.

In the third step, some Z value *must* be programmed. If Z-O is programmed, a 0.100-inch movement will still occur for this type of format. Further, the Z axis value becomes an absolute dimension when programmed in the G79 cycle. The G79 cycle is now used only for basic milling cycles. G01 should be used for more elaborate operations.

The R plane is a reference plane, established by the programmer, which determines where the forward rapid movement of the Z axis terminates. When programming a G79 fixed cycle, however, the R plane is a reference plane that determines where the first forward feed movement of the Z axis terminates. The programmed R plane word usually results in the tool stopping at gage height; however, this is at the discretion of the programmer and conditions surrounding the job. For the example to follow, an R 10.0000 plane has been selected arbitrarily.

Referring to the sample program in figure 9-15, in sequence number O15 the basic cycle G79 is used to move simultaneously the X and Y axes to Pos. 1 from the previous position. This motion will be at the programmed feed rate. When Pos. 1 is reached, the Z axis will feed 0.250 inch to the Z

O/N SEQ.	G PREP. FUNCT.	X ± POSITION	Y ± POSITION	Z ± POSITION	R ± POSITION	I/J/K POSITION	A/B/C P/Q ± WORD POSITION	F/E FEED RATE	S SPINDLE SPEED	D WORD	T TOOL WORD	M MISC. FUNCT.
Ø 15	G79	X 40000	Y 40000	Z- 1500	R 100000		B 0	F 80	S 350		T3	M03
N 16			Y 20000									
N 17	G80											

FIGURE 9-15
G79 — Sample program

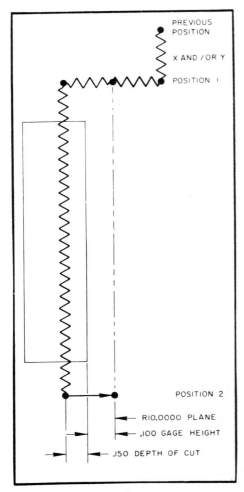

PREVIOUS
POSITION

X AND / OR Y

POSITION 1

POSITION 2

R10.0000 PLANE

.100 GAGE HEIGHT

.150 DEPTH OF CUT

FIGURE 9-16
G79 — Basic mill cycle — cycle representation

depth of 0.150 inch, also at the programmed feed rate. After reaching depth, the next block of information (N16) is read and acted upon.

In sequence N16, the G79 cycle is used again to move the machine slide at the same programmed feed rate to the new Y coordinate (Pos.2) with a 0.150-inch depth of cut. When Pos. 2 is reached, sequence N17 is read. A schematic representation is shown in figure 9-16.

In N17, the G80 R 10.0000 combination will retract the Z axis to the R 10.0000 plane.

BORING

Boring is one of the most accurate ways to finish a hole. When starting with a drilled hole, the sequence of operations usually is to semifinish and finish bore. Starting with a cored hole, the operations generally required are rough, semifinish, and finish bore. Better boring will be achieved if the programmer makes sure that:

- the largest boring bar that will fit the hole to be machined is used.

- a chamfer tool is used, instead of a tool with a square shoulder, whenever possible.

- multiple tool bars are used. The cutting operation should be planned so that the front cutter is through the work before the succeeding cutters start. This is because chatter from one cutter can be transmitted through the bar to the remaining cutters.

- contour milling is employed, whenever practical, because it is often possible to eliminate the rough and semifinish operation by contour milling.

The following axis movements will occur when a G85 bore cycle is programmed:
    1) rapid in X and/or Y.
    2) rapid in Z axis to gage height.
    3) feed in the Z axis to the Z depth.
    4) feed retract to gage height.
These four steps will occur in the same order every time a G85 cycle is programmed. Figures 9-17, 9-18, and 9-19 illustrate the use of a G85 BORE cycle.

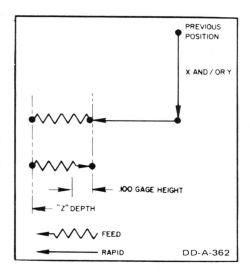

FIGURE 9-17
G85 — Fixed bore cycle schematic

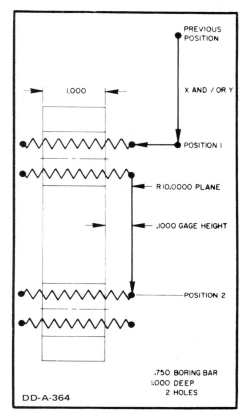

FIGURE 9-18
G85 — Fixed cycle bore — cycle representation

In sequence number O15, the G85 code rapid advances the X and/or Y axes simultaneously to Pos.1 from the previous position. When Pos. 1 is reached, the Z axis will rapid to the R 10.0000 plane (gage height). At this point the Z axis will feed to the programmed depth at the programmed feed rate. After reaching depth, the Z axis will feed retract to the R plane (gage height), and the next block of information (N16) will be read and acted upon.

In sequence N16, the G85 code is used again to rapid the Y axis, with tool at gage height, to Pos.2. Then the Z axis will feed to the programmed

O/N SEQ.	G PREP. FUNCT	X ± POSITION	Y ± POSITION	Z ± POSITION	R ± POSITION	I/J/K POSITION	A/B/C POSITION P/Q ± WORD	F/E FEED RATE	S SPINDLE SPEED	D WORD	T TOOL WORD	M MISC. FUNCT
Ø 15	G85	X+ 40000	Y+ 100000	Z− 10500	R+ 100000		B 0	F 40	S 623		T 9	M02
N 16			Y 80000									

FIGURE 9-19
G85 — Sample program

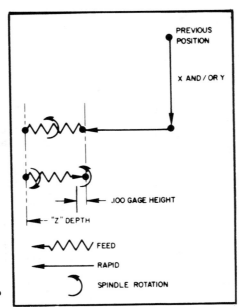

PREVIOUS POSITION

X AND / OR Y

.100 GAGE HEIGHT

"Z" DEPTH

FEED

RAPID

SPINDLE ROTATION

FIGURE 9-20
G84 — Fixed cycle tap schematic

depth at the programmed rate. After reaching depth, the Z axis will feed retract to the R plane (gage height).

TAPPING

When programming tapping operations, be sure that the proper drill has been specified or that the bored hole size is correct. All taps should have adequate clearance to provide *chip* disposal. Avoid using straight-fluted hand or machine taps except when tapping a material such as cast iron. Chips can "ball up" during entry. This may break the tap during "backout." Spiral-fluted taps are better, especially for blind or deep holes, as the spiral causes the chips to feed up the length of the tap and out the hole. For through holes no longer than twice the tap diameter, a straight flute, spiral point, or "gun" tap can be used. This tap has a negative lead ground on the start, or chamfer end, that causes chips to be thrown ahead of the tap. At reversal, it leaves the chips there. Chip difficulties can be eliminated in some cases, by using fluteless taps which roll or form the thread into the walls of the hole. In summary, use gun taps for short through holes, spiral-fluted taps for deep or blind holes, and fluteless taps whenever possible.

The following axis movements will occur when a G84 tap cycle is programmed:

    1) rapid in X and/or Y.
    2) rapid the Z axis to gage height.
    3) feed in the Z axis to the Z depth.

O/N SEQ.	G PREP. FUNCT.	X± POSITION	Y± POSITION	Z± POSITION	R± POSITION	I/J/K POSITION	A/B/C POSITION P/Q ± WORD	F/E FEED RATE	S SPINDLE SPEED	D WORD	T TOOL WORD	M MISC. FUNCT.
Ø 15	G84	X+ 40000	Y+ 100000	Z− 8470	R+ 100000		B 0	F 300	S 700		T 15	M03
N 16			Y 80000									

FIGURE 9-21
G84 — Sample program

4) reverse spindle direction of rotation, and feed retract the Z axis to gage height.
5) reverse the spindle again at gage height.

These five steps will occur in the same order every time a G84 cycle is programmed.

NOTE: *If a right-hand thread is to be tapped, the M function should be for a CW (clockwise) spindle rotation. If a left-hand thread is to be tapped, the M function should be for a CCW (counterclockwise) spindle rotation.*

Figures 9-20, 9-21 and 9-22 illustrate the use of a G84 tap cycle. In sequence number O15, the G84 code rapid advances the X and/or Y axes simultaneously to Pos. 1 from the previous position. When Pos. 1 is reached, the Z axis will rapid to the R 10.0000 plane (gage height). At this point, the Z axis will feed to the programmed depth* at the programmed rate, with the spindle rotating in its primary direction as directed by the M function. At depth, the spindle will reverse direction of rotation and feed retract to the R 10.0000 plane (gage height). Spindle rotation will reverse again to the primary direction. Now the next block of information (N16) will be read and acted upon.

---

*The programmed depth for the Z axis in this example was calculated as follows:

$$Z \text{ position depth to be tapped} + \left[\text{tap chamfer} \times \frac{1}{\text{Pitch}} \right] - \left[\text{Revolutions for reversal} \times \frac{1}{\text{Pitch}} \right]$$

This example is based on a 1/4-20 tap with 3-thread chamfer. Pitch is 20.

Therefore: $\frac{1}{\text{Pitch}} = \frac{1}{20} = 0.050$ inch

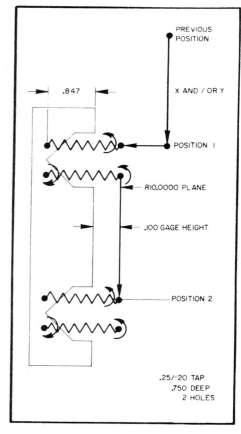

FIGURE 9-22
G84 — Fixed cycle tap — cycle representation

In sequence N16, the G84 code is used again to rapid the Y axis with the tool at gage height, to Pos. 2. Then the Z axis will feed to the programmed depth* as before. At depth, the spindle will reverse direction of rotation and feed retract to the R 10.0000 plane (gage height). Spindle rotation will reverse again back to the primary direction.

OTHER FUNCTIONS

Most modern machining centers are equipped with controls that will perform many additional operations other than the basic cycles already discussed. Some of the more common preparatory and miscellaneous functions follow.

- G00 — Positioning at rapid traverse
- G01 — Linear interpolation*
- G02 — Circular interpolation, CW direction
- G03 — Circular interpolation, CCW direction
- G04 — Dwell
- G17 — Circular in XY plane*
- G18 — Circular in XZ plane
- G19 — Circular in YZ plane
- G79 — Fixed cycle — mill
- G80 — Fixed cycle — cancel
- G81 — Fixed cycle — drill
- G82 — Fixed cycle — drill, dwell
- G84 — Fixed cycle — tap
- G85 — Fixed cycle — bore in, bore out
- G86 — Fixed cycle — bore in, stop spindle, rapid out
- G89 — Fixed cycle — bore in, dwell, bore out
- G90 — Absolute positioning*
- G91 — Incremental positioning
- G92 — Preload absolute stores
- G93 — 1/time feed rate mode
- G94 — IPM feed rate mode*
- M00 — Program stop
- M01 — Optional stop
- M02 — End of program
- M03 — Spindle ON-CW
- M04 — Spindle ON-CCW
- M05 — Spindle OFF
- M06 — Tool change
- M07 — Mist coolant ON
- M08 — Flood coolant ON
- M09 — Coolant OFF
- M13 — Spindle ON CW and flood coolant ON

*These items are usually set up in the control at turn-on, by a data reset, and at the end of each program.

- M14 — Spindle ON CCW and flood coolant ON
- M17 — Spindle ON CW and mist coolant ON
- M18 — Spindle ON CCW and mist coolant ON
- M30 — End of tape

In addition, most modern machining centers and controls are equipped with a variety of options. Some of the more common options follow. (Some are standard, depending upon manufacturer.)

- CDC (cutter diameter compensation)
- Inch/metric switchable input/output
- Automatic interpretation BCD/ASCII tape format
- Position set
- Block delete
- Axis inversion
- Helical interpolation
- Tool gage interface
- Automatic acceleration and deceleration (ACC/DEC)
- Axis error compensation
- Automatic backlash compensation
- Punched tape entry of tool data
- Assignable tool length trims
- Single- or multiple-part program storage
- Tape-punch unit
- Customer-oriented diagnostics

One recent innovation which deserves special attention is a device called "probe" or precision surface sensing, figure 9-23. This device is a creative and time-saving method of accomplishing work centering. Probe is used to electronically trigger the programmed reference in X, Y, or Z and make automatic compensation for the measured axis values through direct feedback.

Essentially, the touch sensor tool can be used to automatically regrid the machine, locate the setup point, establish tool clearance planes, or determine a reference point for machining with respect to the center of a boss or a cored hole.

The surface sensing system contains three parts. The first is the probe body with an interchangeable stylus. (The stylus makes physical contact with the workpiece.) The second part is the noncontacting inductive *module*. (One module is mounted on the nose of the machine spindle while another is mounted on the probe itself.) The third component is the control and interface printed circuit board.

The precision surface sensing probe can be loaded and automatically selected from the tool storage matrix of machining centers, as shown in figure 9-24. It minimizes setup and alignment time by reducing manual operations. In addition, it improves productivity because it centrally locates the finished part within the envelope of the rough part, ensuring part clean-up. The result is a significant, immediate reduction in the time wasted machining parts that do not have sufficient stock. There is also an immediate

FIGURE 9-23
Probe precision surface sensing tool (Courtesy of Cincinnati Milacron Inc.)

FIGURE 9-24
The surface sensing probe loaded in the tool storage matrix of a horizontal machining center (Courtesy of Cincinnati Milacron Inc.)

reduction in the time required to lay out parts manually by bluing and line scribing.

EXAMPLE PROGRAMS

The following example programs are typical of parts which require various tooling to perform some common machining operations. Only certain excerpts have been included to avoid lengthy lists of repetitive operations.

Undoubtedly, there are approaches and proprietary techniques other than those detailed. However, these examples should serve to better acquaint you with the basic functions and operations of modern machining centers.

The first program is for a vertical machining center. The operations required are to spot drill, drill, and bore the 1-inch holes in the part shown in figure 9-25. The program in figure 9-26, while basically very simple, illustrates decimal point programming for a modern vertical machining center. The block by block explanation helps explain the programming used to complete the necessary machining.

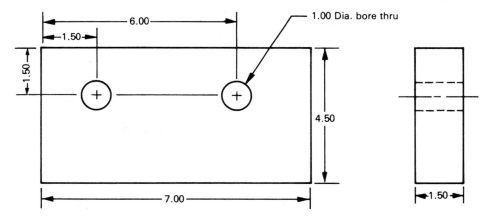

FIGURE 9-25
Vertical machining center programming example part

Program	Explanation
O6001	Program part name
N1 G92 X0 Y0 Z0 T01	Load absolute preset values X = 0, Y = 0, Z = 0 (spot drill) (0,0) upper left of part. Tool drum rotate to tool #1
N2 G00 G90 S1100 M03	G00-Rapid, G90-absolute input, 1100 rpm, spindle on CLW
N3 M06 T00	Tool change — discard empty canister
N4 X 1.0 Y-1.5	Rapid to first position
N5 G45 Z.1 H01 T02 M08	Select tool 2 (15/16 drill) G45/H01 tool length offset Rapid to .100 on part surface (Z.1) coolant on (.1108)
N6 G81 R.1 Z-.343 F11.0	Std. drill cycle, .100 above part surf. .343 DP. 11.0 IPM
N7 X6.0	Rapid to and complete 2nd hole
N8 G80 G28 Z.1 M06 T01	G80-cancel G81, G28 Z.1 M06 T01. rapid to home pos. change tool
N9 G00 S850 M03	Rapid traverse, 850 rpm, spindle on CLW.
N10 G45 Z.1 H02 T03	Select tool 3 (1.00 boring bar) G45/H02 tool length offset rapid to .100 of part surface (Z.1)
N11 G83 R.1 Z-1.862 Q.5	Deep hole cycle — G83 peck drill rapid to .100 of part surface peck drill in .500 increments to Z-1.862
N12 X1.0	Rapid to X1.0 and repeat cycle
N13 G80 G28 Z.1 M06 T02	G80-cancel G81, G28 Z.1 M06 T02-rapid to home pos. change tool
N14 G00 S1000 M03	Rapid traverse to position 1000 rpm, spindle on CLW
N15 G45 Z.1 H03 T00	Select T00 (empty cannister) G45/H03 tool length offset rapid to .100 of part surface (Z.1)
N16 G76 R.1 Q0.10 Z-1.562 PO	Finish bore — rapid Z to R. 100 of part surface Po-dwell spindle stop and orient (shift X out .100) Q0.10
N17 X6.0	Rapid to X6.0 and repeat cycle
N18 G80 G28 Z.1 M06 T03	G80- cancel G81, G28 Z.1 M06 T02. rapid to home pos. change tool
N19 G28 Y0	Move to part offload position
N20 M00	Program stop
N21 G00 X0 Y0	Rapid back to program start point
N22 M30	Tape rewind — end of program

FIGURE 9-26
Example program for a decimal point programmed vertical machining center. This program, with block by block explanation, is for the sample part shown in figure 9-25.

The second program is more complex. It is programmed for a modern horizontal machining center using the *interchangeable variable block format.* The following information is provided:

1) An engineering drawing of a hydraulic pump housing, figure 9-27.
2) A plan view, figure 9-28, and an elevation view, figure 9-29, of the one-station indexing fixture. This fixture would permit machining on four sides of the part.
3) The beginning of the N/C program, figure 9-30, along with a detailed explanation for this part of the program.
4) Figure 9-31 illustrates numbered hole positions on the 0° side of the part. Specific blocks of information for drilling the numbered hole positions have been omitted because of their repetitive nature.
5) Figure 9-32 includes blocks of information and explanation for a rotary table index from the 0° to the 180° side of the part for the remaining machining operations.

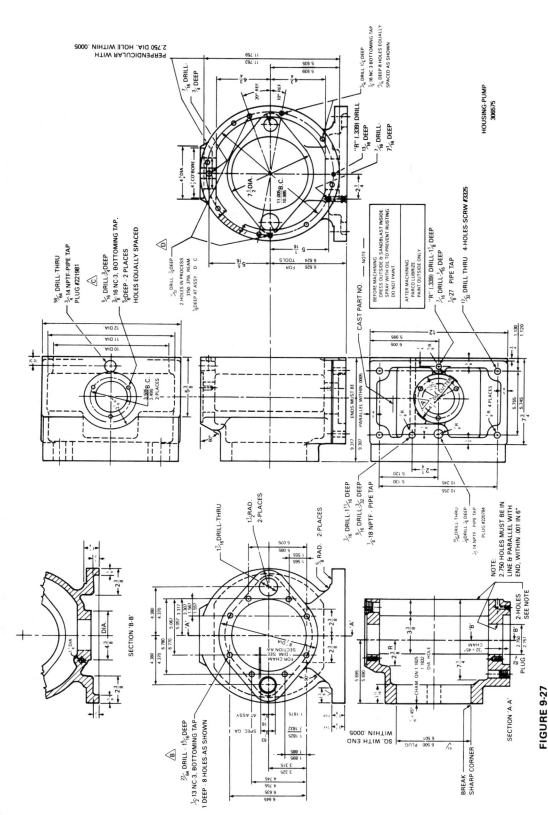

FIGURE 9-27

Sample part drawing (Courtesy of Cincinnati Milacron Inc.)

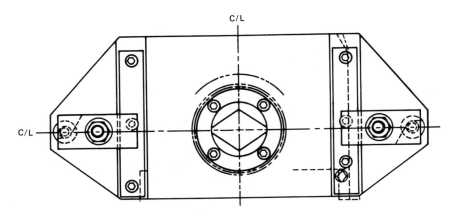

FIGURE 9-28
Fixture plan view (Courtesy of Cincinnati Milacron Inc.)

6) Figure 9-33 includes blocks of information and a related description for drilling and tapping hole positions illustrated in figure 9-34.

7) The blocks of information and tape sequence explanation in figure 9-35 provide the programming information required to profile mill a cored hole to a rough bore as shown in figure 9-36.

8) A tool instruction sheet is shown in figure 9-37 with complete tool identification and cutting statistics.

9) Figures 9-38 and 9-39 illustrate examples of tool drawings to document the various parts of the entire tool assembly and complete the programming support paperwork.

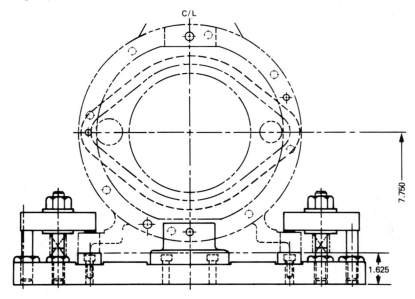

FIGURE 9-29
Fixture elevation view (Courtesy of Cincinnati Milacron Inc.)

O/N SEQ.	G PREP FUNCT	X± POSITION	Y± POSITION	Z± POSITION	R± POSITION	I/J/K POSITION	A/B/C POSITION P/Q ± WORD	F/E FEED RATE	S SPINDLE SPEED	D/H WORD	T TOOL WORD	M MISC. FUNCT
0 1	G00										T 1	M 6
0 2	G00	X 75000	Y 77500	Z 56925			B 0		S 350		T 2	M 3

Tape Sequence:

(E.O.B.)

O1   Tape alignment blocks are identified by the alphabetic character "O" as per E.I.A. standards. With tool changing machining centers, double O blocks are used so that realignment can be accomplished without having to perform a tool change if the correct tool is already in the spindle.

Loading the first tool is accomplished by tape sequence O1. The first tool change will automatically cause the Z slide to go to full retract position prior to loading the spindle with tool number T1.

O2   Illustrates the first slide movement at rapid traverse rate (G00) from the random X and Y location where the first tool was loaded. The resultant tool tip path is a **straight line** between the two points in the XY plane, followed by the Z-axis movement. (G00 mode causes X-, Y-, and B- axes movements to occur simultaneously, followed by Z movement.)

This move locates the cutter at the start of the first cut.

Block number two provides the starting coordinates for all axes (X, Y, Z and B), also the mode of operation (G00), spindle speed (S___), spindle (On/off and direction), coolant (On/off and type) (M___) and number of the next tool used (T___). With one exception, any (format) word left out of this block will still contain the last value present in the N/C control memory and the machine will respond according to old data.

Start-up of the control will automatically assume the following functions are in effect:
   G01 Linear Interpolation
   G17 Circular Plane Selection in XY Plane
   G90 Absolute Positioning
   G94 Inches per minute feed rate
The O address permits tape search to locate this block for realignment purposes. For example, if after executing several blocks of tape it should be desired to start over at the beginning, the operator would search block O2 rather than O1 because the first tool is already in the spindle and re-aligning at O1 would cause an undesired tool change.

FIGURE 9-30

Sample program and explanation for first tool change (Courtesy of Cincinnati Milacron Inc.)

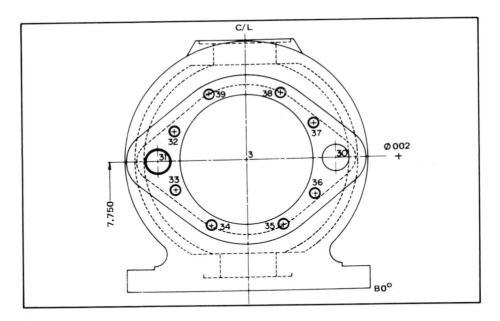

FIGURE 9-31
Numbered hole positions for 0° side of the sample (Courtesy of Cincinnati Milacron Inc.)

O/N SEQ.	G PREP. FUNCT.	X± POSITION	Y± POSITION	Z± POSITION	R± POSITION	I/J/K POSITION	A/B/C POSITION P/Q ± WORD	F/E FEED RATE	S SPINDLE SPEED	D/H WORD	T TOOL WORD	M MISC. FUNCT.
N 16		X 0	Y 40000									
N 17	G00			Z 100000								
N 18				Z 36195			B 180000					
N 19	G01	X 50000						F120				

Tape Sequence

N16   Shows slides in their last cutting motion before index.

N17   X- and Y-axes remain at their last cutting position and Z-axis moves at Rapid Traverse rate to a clearance plane which clears any possible interference between cutter and fixture or workpiece.

N18   B-axis rotates to 180 degrees then Z-axis rapids to depth to start milling.

N19   X-axis starts to feed at programmed feed rate.

FIGURE 9-32
Sample program for index of rotary table (Courtesy of Cincinnati Milacron Inc.)

O/N SEQ.	G PREP. FUNCT.	X± POSITION	Y± POSITION	Z± POSITION	R± POSITION	I/J/K POSITION	A/B/C P/Q	POSITION ± WORD	F/E FEED RATE	S SPINDLE SPEED	D/H WORD	T TOOL WORD	M MISC. FUNCT.
O182	G00											T16	M 6
O183	G81	X 54650	Y 67949	Z -12500	R 36195		B 180000		F50	S1100		T17	M 3
N184		X 31467	Y 32447										
N185		X- 9551	Y 23335										
N186		X-50553	Y 45953										
N187		X-54165	Y 87051										
N188		X-35147	Y 122553										
N189		X 9551	Y 131665										
N190		X 45053	Y 109047										
N191	G00			Z36295									
O192	G00											T17	M 6
O193	G84	X 45053	Y 109047	Z- 9375	R 36195		B180000		F254	S 440		T19	M 3
N194		X 54650	Y 67949										
N195		X 31467	Y 32447										
N196		X- 9551	Y 23335										
N197		X-50553	Y 45953										
N198		X-54165	Y 87051										
N199		X-35147	Y 122553										
N200		X 9551	Y 131665										
N201	G00			Z36295									
O002	G00											T19	M 6

FIGURE 9-33
Sample program for drilling and tapping an eight-hole pattern (Courtesy of Cincinnati Milacron Inc.)

Tape Sequence:

O182   Loads tool (T16) 5/16 x 7/16 subland drill and cosink in the spindle.

O183   G81 (Drill Cycle) causes X-, Y- and B-axes to position at hole 12, first hole of the 8 hole pattern. At the same time the spindle also changes speeds to 1100 rpm in the CW direction (M03). Then Z-axis moves at rapid traverse to the R plane position, feeds at 5.0 ipm (F50) to a depth of 1.2500 inches (Z-12500), and then rapid retracts to the R plane ready to start next operation.

N184   Drills the remaining 7 holes. (13 thru 19). These are the same as the
thru   first hole, except for the X, Y locations, so only the X, Y values are
190   programmed in these blocks.

N191   Retracts the Z-axis to gage height.

O192   Loads (T17) 3/8 - 16 NC tap in the spindle and returns tool (T16) to its proper location in the matrix.

O193   G84 (Tap Cycle) taps hole P19 first of the 8 hole pattern. The slides do not move since they were already in proper position for this hole. The spindle will change speeds to 440 rpm in the CW direction (M03). Then the Z-axis moves at rapid traverse to the R plane position, and feeds at 25.4 ipm (F254) to a depth of 0.9375 inches (Z-9375). At depth the spindle reverses and the Z-axis feeds back to the R plane ready to start next operation.

N194   Taps the remaining 7 holes. These are the same as the first tapped
thru   hole, except for the X, Y location, so only the new X, Y values are
200   programmed in these blocks.

N201   Retracts the Z-axis to gage height.

O202   Tool (T19) is loaded into the spindle for the next operation and T18 returned to storage.

FIGURE 9-33
Continued

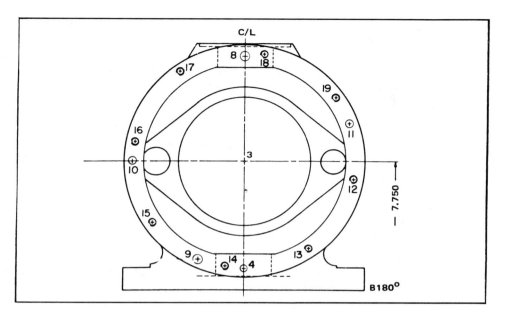

FIGURE 9-34
Sample workpiece (Courtesy of Cincinnati Milacron Inc.)

O/N SEQ	G PREP FUNCT	X± POSITION	Y± POSITION	Z± POSITION	R± POSITION	I/J/K POSITION		A/B/C POSITION P/Q ± WORD		F/E FEED RATE	S SPINDLE SPEED	D/H WORD	T TOOL WORD	M MISC. FUNCT.
Ø216	G00												T 2	M 6
Ø217	G00	X 0	Y 77500	Z 56925				B 0			S 555		T 3	M 3
N218	G01			Z 49000						F500				
N219	G02	X-11050	Y 66450			I-11050	J 77500	P 0	Q 10000	F250				
N220		X-22100	Y 77500					P 10000	Q 0	F 76				
N221		X 0	Y 99600			I 0		P 0	Q-10000					
N222		X 22100	Y 77500					P-10000	Q 0					
N223		X 0	Y 55400					P 0	Q 10000					
N224		X-22100	Y 77500					P 10000	Q 0					
N225		X-11050	Y 88550			I-11050		P 0	Q-10000	F250				
N226		X 0	Y 77500							F500				
N227	G01			Z 39425										

FIGURE 9-35
Sample program for circular milling with CDC (Courtesy of Cincinnati Milacron Inc.)

A 2 inch diameter end mill is used to circle mill a cored hole to a rough bore dimension of 6.4200″, thus eliminating the need to have a boring bar set to this dimension. Since the 2 inch diameter end mill is used elsewhere in the program this technique saves a space in the tool matrix. This technique is valuable when stock removal is heavy or irregular, but especially when the tool matrix is full and no more tools can be added.

The cutter approach path is a semi-circle tangent to the 6.42 inch diameter circle rather than a straight line. This brings the cutter gradually into contact and eliminates "Wrap around" which could set up chatter because of the large arc of contact.

Tape Sequence:

O216   Loads 2 inch end mill (T02).

O217   Positions the cutter at the center of the cored hole with CDC off, and rapids Z-axis to gage height. Selects the proper spindle speed and starts the spindle in the CW direction.

N218   Z-axis feeds to depth at 50 ipm.

N219, 220   G02 produces a clockwise semicircular approach path tangent to the 6.50 diameter hole. I and J coordinates define which of the two circles is being used. P and Q coordinates cause the system to offset the cutter from the program path by one-half the CDC value stored in the control memory for this particular tool.

N221, 2,3,4   Each block produces a 90 degree arc of the circle at a feed rate of 7.6 ipm. The feed rate at the periphery is 12 ipm. The new I and J values cause the cutter to follow a new circular path after reaching the tangency point. P and Q values perpetuate the CDC offset already in effect. The vector values are defined at the end of each circular span — intermediate values are calculated by the control.

N225   I and J are programmed here to cause the cut path to follow a semicircular exit path away from the work. Feed rate is increased since no more metal is being removed. CDC remains "ON" until the cutter is away from the work, to eliminate any marks on the work which might result from cutter deflection.

N226   No P and Q values are present so CDC offset is reduced to zero and the cutter feed rate is stepped up to 50 ipm. Final position is at the center of the cored hole at span end.

N227   Z-axis moves to a new setting for the next cut.

FIGURE 9-35
Continued

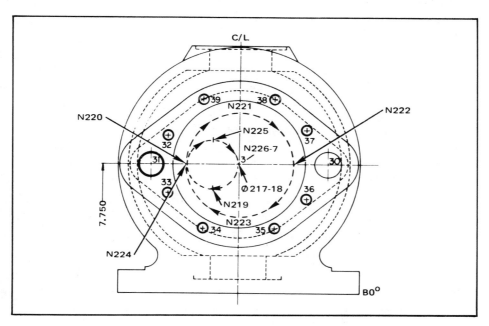

FIGURE 9-36
Sample part — milling (Courtesy of Cincinnati Milacron Inc.)

PART NAME			PART NO.			DRAWING NO.	
HOUSING — PUMP VARIABLE DELIVERY			306575				
MACHINE AND CONTROL			REVISION	PROG. BY		DATE	
CIM-X CHANGER 720 with CNC CONTROL						5/8/84	
SETUP AND TOOL INFORMATION			CHECK BY			PAGE OF PAGES	
FIXTURE: USE 2CB-6 8M-434163 SETUP: USE 0.100 IN. GAGE						1	1
TO SET TLC TRAM POSITION X = 0.0000, Y = 3.7500, Z = 6.1000, B = 0							

OPERATION or STATION NUMBER	TOOL			REMARKS	CUT SPEED	R.P.M.	FEED	CPT or FD/REV
	DESCRIPTION	ASSEMBLY No.	SET DIM.					
1	3" CARBIDE END MILL	10011	3.5		300	382	23.0	.052
2	2" CARBIDE END MILL	11030	8.00		224	466	20	.043
3	17/32 DRILL	01022	5.625		80	575	4.9	.008
4	.520 X .5315 BORESIZE DRILL (CARB.)	99015	6.94		250	1800	30	.0167
5	1-1/8 X 1-1/4 SUBLAND DRILL	02014	7.9		82	250	3.0	.012
6	59/64 DRILL	01025	7.4		60	250	3.0	.012
7	3/4 — 14 NPTF PIPE TAP	08006	8.0		40	145	9.4	.065
8	7/16 X 9/16 SUBLAND DRILL	02011	8.0		80	700	5.0	.007
9	1/4 — 18 NPTF PIPE TAP	08009	6.19		40	230	12.3	.056
10	45/64 X 7/8 SUBLAND DRILL	02010	6.9		70	380	4.0	.011
11	1/2 — 14 NPTF PIPE TAP	08005	6.19		40	182	12.5	.069
12	R (.339) X 7/16 SUBLAND DRILL	02009	5.25		80	890	5.4	.006
13	1/8 — 27 NPTF PIPE TAP	08004	4.75		40	377	14.0	.037
14	2.740 BORE AND CHAMFER BAR	09042	6.90		300	416	2.2	.005
15	2.750 BORING BAR	09017	6.75		300	416	2.4	.006
16	5/16 X 7/16 SUBLAND DRILL	02013	5.25		80	977	5.0	.005
17	3/8 — 16 NC TAP	07012	5.56		40	406	25.4	.0625
18	1-7/16 DRILL	01024	8.41		60	158	1.1	.007
19	7/16 DRILL	01020	10.5		60	525	3.1	.006
20	11/32 DRILL	01021	4.88		80	878	5.3	.006
21	27/64 X 9/16 SUBLAND DRILL	02015	5.75		80	725	5.8	.008
22	1/2 — 13 NC TAP	07013	5.94		40	305	23.5	.077
23	1.1725 BORING BAR	09015	6.25		247	805	6.0	.0075
24	6.5000 BORING BAR	09016	3.375		303	176	1.1	.006

FIGURE 9-37
Sample tooling form

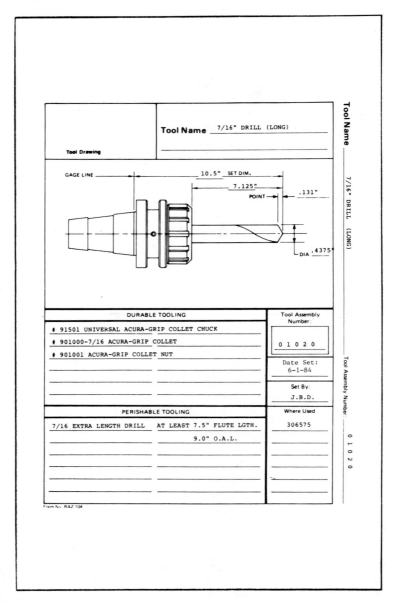

FIGURE 9-38
Tool assembly drawing

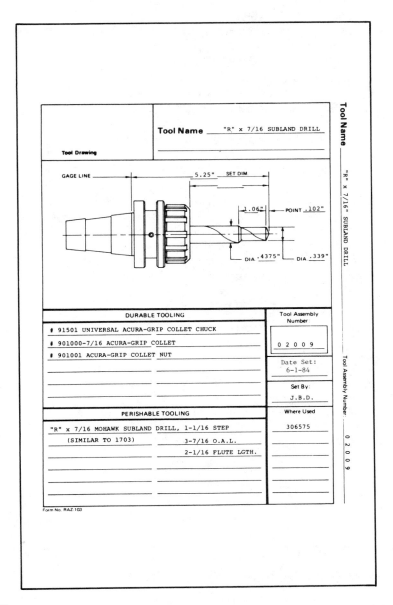

Tool Name   "R" x 7/16 SUBLAND DRILL

Tool Drawing

GAGE LINE   5.25"   SET DIM

1.06"   POINT .102"

DIA .4375"   DIA .339"

DURABLE TOOLING

91501 UNIVERSAL ACURA-GRIP COLLET CHUCK

901000-7/16 ACURA-GRIP COLLET

901001 ACURA-GRIP COLLET NUT

Tool Assembly Number

0 2 0 0 9

Date Set:
6-1-84

Set By:
J.B.D.

Where Used

306575

PERISHABLE TOOLING

"R" x 7/16 MOHAWK SUBLAND DRILL, 1-1/16 STEP

     (SIMILAR TO 1703)   3-7/16 O.A.L.

                         2-1/16 FLUTE LGTH.

Form No. RAZ-103

Tool Name   "R" x 7/16" SUBLAND DRILL   Tool Assembly Number: 0 2 0 0 9

FIGURE 9-39
Tool assembly drawing

REVIEW QUESTIONS

1. List the major elements which contribute to the versatility of a machining center. Briefly explain their functions.
2. How are variations in tool lengths handled on a machining center?
3. What types of safety precautions must be considered prior to any tool change and/or table index?
4. What are tool assembly drawings? Why are they used?
5. What types of conventional and unconventional operations are performed on machining centers?
6. What kinds of indexing options are available on modern machining centers?
7. Explain the difference between random and sequential tooling.
8. Name some machining center format words. Explain their functions.
9. Why are drill points calculated when programming for drilling operations?
10. Name some optional features available on machining centers.
11. What is the primary function of the probe surface sensing tool?

CHAPTER 10

Numerical Control Programming with Computers

OBJECTIVES
After studying this chapter, the student will be able to:

- Understand the importance of computers in numerical control applications.
- Name the various computer languages available for numerical control, and describe their general characteristics.
- Discuss the general format and capabilities of the APT language.
- Identify the major functions of a postprocessor.
- Explain the basic differences between hard-wired and soft-wired controls.
- Understand DNC (direct numerical control) and its impact on numerical control.
- Describe the general features and functions of a CNC unit.

The computer has taken over many facets of work normally done manually. Unfortunately, many people have been given the wrong impression about the computer and the way it functions. Media leads us to conclude that the computer is some kind of magic brain that sees, hears, and knows all. In reality, a computer is simply a tool that will perform a given task, providing the computer and man communicate in the same language. Computers can save hundreds of work hours and can process data more economically and accurately.

It is difficult, however, to convince some people that users of computers do not necessarily need a highly technical background in order to profit from their use. Only an understanding of the language is required, and one can make the computer a slave for the task one wishes it to perform.

CHARACTERISTICS OF COMPUTERS

Computers are very complex instruments. However, they basically perform three functions: accept data; process data; and develop an output.

Speed is a very important factor in accomplishing these three jobs. Computers can perform many repetitious functions in only a fraction of the time it would take a person. For example, many computers can add a quarter of a million numbers having sixteen digits in one second!

Another important characteristic of computers is the amount of memory or storage capacity they possess. The basic unit of logic or *storage* is the word composed of *bits* which are charged or discharged "cores" within the memory of the computer. Accuracy is also a key element in working with computers. Since people, not computers, make errors, the accuracy of the computer is limited only by the accuracy of the programmer.

This then brings up the real key to the profitable use of computers — programming. A computer is completely without initiative. It follows whatever set of instructions it has been given, and is helpless when confronted with a situation that was not specifically covered in its instructions. Therefore, every conceivable eventuality must be allowed for, and suitable instructions must be provided. These masses of electronic components and circuitry could not add one and one without being given a set of explicit instructions.

N/C AND COMPUTERS

Some companies, having only a few N/C machines, find that they need computer assistance for some of their work. If they do not own a computer, they are able to rent time or hire a specialist who has knowledge of computer programming and access to a computer.

Using the computer to prepare tapes for numerically controlled machines greatly reduces the cost of tape preparation, particularly if the part is complex. It also produces accurate tapes more often than manual programming. The computer-aided part programming system will also provide error diagnostics relating to format, spelling, and typographical errors at the computer level where it is generally cheaper to correct, rather than having errors detected at the machine tool. This reduces the amount of machine time that might normally be wasted because of tape or programming errors. Figure 10-1 illustrates the flow of an N/C program when using a computer.

Most large companies who depend on computer support for providing their N/C tapes usually communicate their N/C services by means of processor languages. Currently, there are dozens of N/C processor languages available. While most organizations that use numerical control make use of one of the existing processor languages, some have actually written their own language for special applications.

These particular processor languages — other non-N/C processor languages include FORTRAN (formula translation), COBOL (common business oriented language), BASIC — take up an extensive amount of computer logic or memory, depending on the amount of storage capacity needed to accommodate a particular processor language. These languages must contain enough capabilities to execute the instructions called for in a manuscript. If the processor has a very limited capability, the programmer must confine the

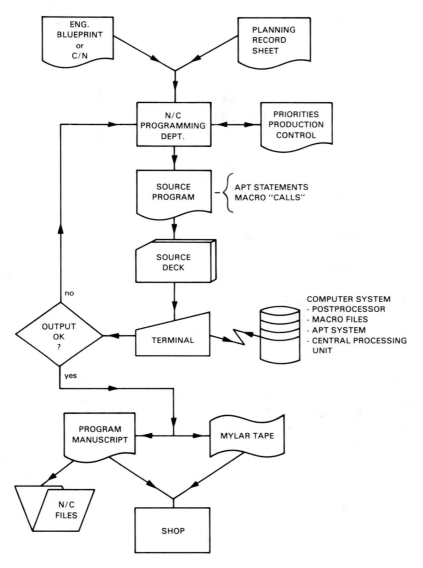

FIGURE 10-1
Flowchart of the development of an N/C program

manuscripts to that capability. The part programmer must understand the terms of the specific language being used and how these terms will be processed by the computer.

Therefore, the more powerful and complete the language, the larger the computer's capabilities. This is why so many processors have been written. It is also why there is a trend toward using a small computer for processing a workpiece program while all processor language storage is done on some auxiliary piece of equipment, such as a disc pack or magnetic tape storage. Simple processors use smaller computers, and the simple languages may be quite adequate for basic N/C work.

COMPUTER LANGUAGES AVAILABLE FOR N/C

As mentioned, due to the large variety of processor or programming languages available, choosing the correct program language is often a very difficult and major decision. Types of parts produced, types of machine tools used, lot sizes, costs, materials used, and computer availability should all be considered when selecting a particular part programming language. The following are the most common types of computer languages.

APT (AUTOMATIC PROGRAMMED TOOL)

This has the largest vocabulary of the general processor languages. It was also the first to be developed. *APT* can be used only on large-capability computers. It can perform the complicated mathematics of complex curves, using four- and five-axis contouring techniques. APT is the basic N/C computer program for the industry. It will be discussed in more detail later.

AD-APT (ADAPTATION OF APT)

This is a limited version of APT and can be run on medium-sized computers. It uses only about one-half of the vocabulary words as APT. *AD-APT* development was initially sponsored by the Air Force. The language is restricted to two simultaneous contouring motions in a plane and a third axis of linear control.

AUTOMAP (AUTOMATIC MACHINING PROGRAM)

This language is a further modified version of APT. It works primarily with straight lines and circles, and it runs on medium-sized computers. AUTOMAP is easy to learn, and uses about fifty vocabulary words as well as some specific punctuation.

COMPACT II

This language and its processor, while used widely throughout industry, operate in an interactive environment on remote, time-shared computers. The programmer prepares a manuscript, and then communicates it to the computer system via an input/output terminal. This language encompasses a broad spectrum of users from some of the largest manufacturing concerns. It can cover N/C applications ranging from the standard milling, drilling, turning, boring, etc. to punch press work, EDM, and flame cutting.

The flexibility of COMPACT II makes it an easy language to learn. Basically, there are only a few rules to memorize. The symbols and language organization are simple and may be learned in a relatively short period of time. The words used in the COMPACT II language are easily recognizable, even without formal training. The words may be placed in any sequence within the programming statement. This allows a free-form use of the language and stems from the language that is basic to it, the SUNDSTRAND SPLIT language. However, statement structure and sequences follow logical N/C machining procedures or steps. This language specifies the machine, defines the part shape, selects the tools, and directs the cutting motion. The computer helps *debug* the input through interactive conversation with the programmer while processing the program, and returns, via the terminal, the machine control tape. The conversational interactive nature of the COMPACT II processor provides step-by-step diagnostics (errors) as the manuscript is transmitted to the remote computer for processing.

UNIAPT

This language and its processor operate in a batch mode on a local dedicated minicomputer. UNIAPT can handle programming for all three axes and most four- and five-axis machine tools. The UNIAPT language is almost identical to the APT language; however, a few words in APT and UNIAPT are not compatible with each other. The major distinction between the two is the size of the computer required. APT requires a large computer system, while UNIAPT can run on a minicomputer.

NUFORM

This language and its processor are used by inserting codes, or dimensional numbers, in appropriate eighty-card columns. The columns are divided among ten fields. Each NUFORM statement conforms to the structural rules of one of about eighty modules. NUFORM uses numerical codes rather than mnemonic codes; letter codes, abbreviations, and punctuations are used rather than words.

SPLIT (SUNDSTRAND PROCESSING LANGUAGE, INTERNALLY TRANSLATED)

This is a proprietary language of Sundstrand and must be run on a large computer. The language and its processor operate in either a batch or interactive mode using a local dedicated or time-shared computer. The programmer writes a manuscript on a ruled form. Each manuscript line is keypunched into cards and fed to the computers. The computer processes the program and output cards. These are listed and may also be output as punched tape for machine control.

The vocabulary used in many of these programs is very similar. When one learns some point-to-point and contouring routines of one language, the transition to any other language should be fairly easy. However, because APT was the first language to be developed and the other languages are derivatives of APT, we will concentrate on APT and its use.

APT GENERAL PROCESSOR

APT stands for automatically programmed tool, and refers to a language and computer program. Currently, there are about fifteen widely used implementations of APT in the United States. Each has its own set of changes or additions to the original APT system. These differences are relatively minor, and do not affect the basic parts of the language.

To better understand the nature and function of APT, a brief discussion of its background is necessary. While APT has an interesting history and is the base from which all other N/C computer languages evolved, it may not be the best choice for *all* applications and situations.

N/C was developed as an answer to some complex aerospace machining problems, such as the aerodynamic curves of blade and air foil surfaces. At that time, each aerospace company tried to write its own processor language. However, the companies found that the job required more time and man-power than could be committed. Finally, the members of the AIA pooled their resources in a cooperative development project. In 1961, it was decided to broaden the scope of APT; further development was turned over to IIT research in Chicago. The APT system soon emerged, representing over 100 years of development and testing. The membership was then opened to include other industries. The APT Long Range Program was further broadened to include *CAM-I* (Computer Aided Manufacturing International).

Basically, the APT system is divided into four sections, plus Section-0. Section-0 may be considered the supervisor or operating system which controls the flow of information in the various sections of the system. This is illustrated in figure 10-2.

In Section-1, the input translator phase reads the source statements one at a time. As each statement is processed, it is checked for errors in punctuation, ordering, incomplete statements, and syntax. Any errors detected in this phase cause an *error signal* to the part programmer. The signal indicates the type of error and the correction procedure to be followed. The source statements are then separated and classified by type of operation. The necessary data is extracted, rearranged, and recorded into computer-processible form for the next phase. Certain data dealing entirely with machine tool functions, e.g., coolant control, which is not required to compute the cutter center point or path is coded through to the next phase of the processor to the second step of the process.

In Section-2, the arithmetic phase of the processor receives the data from the input translator phase. Using a built-in library of subroutines, tables, and symbols, this phase generates the equations which describe a

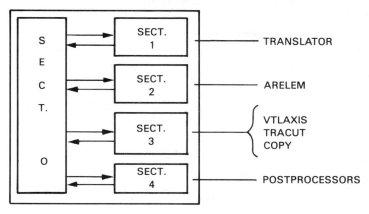

FIGURE 10-2
Sections of APT programming

given machining problem. For example, the problem in question might be the intersection of two lines or arcs, the point of tangency of a line and an arc, or the points which describe a circle to be segmented into straight-line cuts within the required tolerance. These equations are solved to find the coordinate values describing the cutting tool's center point in three-dimensional space. These values are formatted into generalized machining instruction sets as the final output of the processor program.

If no errors are found in Section-2, control is passed to Section-3, the *edit* phase. There are three major functions of Section-3. One is vertical tool axis (a variable tool axis) which is a multiaxis control dealing with the orientation of the spindle from the vertical position. The others, TRACUT and COPY, are used to transform and manipulate the output data of Section-2. If no errors are found in Section-3, or if there is nothing to be done in this section, control is passed on to Section-4.

APT is a multipass processor. It completely processes the input, treating only some aspects (such as spelling and punctuation checks) before another aspect, such as calculations of cutter positions, is processed.

Section-4 is the *postprocessor.* The proper postprocessor is selected from the instructions on the MACHIN/card. The data is converted into the proper format for the specific machine tool which is called out.

POSTPROCESSORS

Postprocessor is the most misunderstood term in numerical control. It has been mistakenly considered a piece of hardware or a separate "black box" sitting off in the corner waiting to postprocess some information.

A postprocessor is a set of computer instructions which transforms tool centerline data into machine motion commands using the proper tape code and format required by a specific machine control system. Also included are feed rate calculations, spindle speeds, and auxiliary function commands.

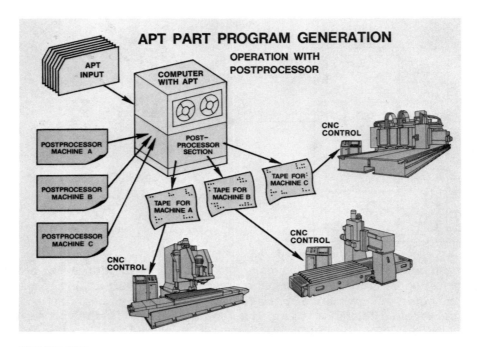

FIGURE 10-3
Relationship of APT program generation, postprocessor, N/C tape, and machine tool (Courtesy of Cincinnati Milacron Inc.)

If APT or any other universal type processor is used, a postprocessor program must be written for each different machine tool/control unit combination that will be used. Most machine tool and control unit builders have developed postprocessors for their own equipment.

Because APT is universal, it cannot convert any calculated data into specific tape formats for any machine tool/control unit. The initial output will be the CL or centerline output. This tells where the centerline of the cutter path is located with respect to the part configuration within the machine coordinate system. An additional step — post processing — is required to adapt the CL output to the particular machine tool/control unit combination which will be used to machine the workpiece. The actual program tape to be used on the machine tool is a product of the postprocessor. This important relationship is illustrated in figure 10-3.

The primary functions of a postprocessor are to:

- convert cutter centerline data to the machine coordinate system.

- ensure that the physical limits of the machine are not exceeded, e.g. range, feed rate.

- contain the part to a given tolerance by controlling the amount of overshoot.

- eliminate reader limitations.

- output preparatory and miscellaneous functions.
- calculate cutter compensation information.
- generate circular or parabolic points.
- generate error diagnostics when necessary.

There are many varieties of postprocessors available. For this reason, a part programmer should study thoroughly the postprocessor documentation in order to become familiar with the capabilities and requirements of the postprocessor.

WRITING AN APT PROGRAM

APT part programming involves three major elements:

- Definition and symbolic naming of geometric points and surfaces representing part size and configuration.
- Specification of cutting tool and action or tool motion statements. (These statements move the cutter to the points or along the defined geometric surfaces.)
- Specifications of conditions required at the machine tool such as spindle speeds, feed rates, and other auxiliary function commands. (These are relative to the independent postprocessor being used.)

A simple APT program is shown in figure 10-4. In providing the geometric definitions of the part, the APT part programmer communicates to

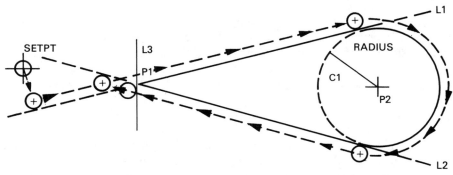

APT GEOMETRY STATEMENTS
SETPT = POINT/X, Y, Z
  P1 = POINT/X, Y, Z
  P2 = POINT/X, Y, Z
  C1 = CIRCLE/CENTER, P2, RADIUS, R
  L1 = LINE/P1, LEFT, TANTO, C1
  L2 = LINE/P1, RIGHT, TANTO, C1
  L3 = LINE/P1, ATANGL, 90
CUTTER/.5

APT MOTION STATEMENTS
FROM/ SETPT
RAPID
GO/ TO, L1
FEDRAT/20
TLLFT, GOLFT/L1, TANTO, C1
GOFWD/C1, TANTO, L2
GOFWD/L2, PAST, L3
GOTO/SETPT
FINI

FIGURE 10-4
Simple APT program

the computer using the specific APT vocabulary. In this vocabulary, there are approximately two hundred and sixty words, including punctuation. The APT geometry statement consists primarily of three parts:

Symbol = surface/description
C1 = Circle/center, P2, radius, R

APT DEFINITION STATEMENTS

The first part is a symbol which is an arbitrary name assigned to a particular geometric element. This symbol is then equated to the definition (the second part of the statement) which is a major word such as point, line, circle, etc. The major word defines the type of surface or geometric element that the symbol represents. The third part of the APT statement is the actual description which consists of minor words or modifiers and numerical values of the point, line, circle, etc. These position the element in space and determine its specific size.

APT MOTION STATEMENTS

The motion statements in the APT language are typical of statements that might be used in directing a person to walk around the block or through town.

Positional modifier	Directional modifier	Drive surface	Modifier	Check surface
TLLFT	GOLFT /	L1	TANTO	C1

A much more detailed, typical, and complete APT program, along with its postprocessor printout, is shown in figure 10-5. Included in this figure are operations sheet and commentary contained in the program.

The APT part programmer must first define the part to be produced and its various elements and surfaces, selecting the best available format from the APT language. There are numerous formats available, and each must be used in exactly the same way it is provided in the APT vocabulary. The part programmer does not have the freedom to invent or modify any APT definition statement.

As computer part programming continues to grow, more companies will be relying upon either APT or one of its derivative languages for support. This can be done through direct purchase of a computer or through time sharing. Additional information can be obtained from any N/C machine manufacturer and technical institutes and colleges teaching the subject.

HARDWARE VERSUS SOFTWARE

When discussing machine control units, many questions and concerns are raised regarding the terms *hardware* and *software*. A greater appreciation

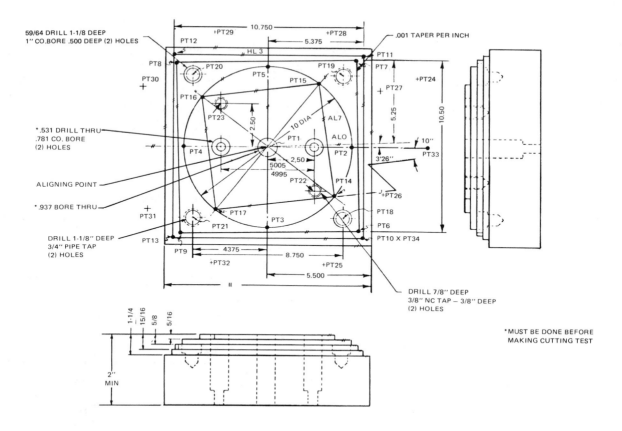

SEQUENCE OF OPERATIONS

1. Align Point — 1" bored hole (X000000 Y000000) P1
2. Mill Off Corners of Part — 5/8 deep, 2 passes, 5/16 stock removal each pass 2" carb. end mill, 8 ipm. P25–P32
3. Mill Circle — 5/8 deep, 2 passes, 5/16 stock removal each pass, 2" carb. end mill, leave .020 on radius for finish cut, 12 ipm. P2–P5
4. Mill Square Inside Circle — 5/16 deep, 1 pass, 2" carb. end mill, square at 10° angle, leave .020 on sides for Finish cut, 8 ipm. P14–P17
5. Mill (Taper) Square — 5/16 depth of cut (15/16 deep from top of part) 1 pass, 2" carb. end mill, leave .020 on sides for Finish cut, 14 ipm. P6–P9
6. Mill Square — 5/16 depth of cut (1-1/4 deep from top of part) 1 pass, 2" carb. end mill, leave .020 on sides for Finish cut, 14 ipm. P10–P13
7. Mill Square To Finish Dimension — 2" carb. end mill, 1 pass 20 ipm. P10–P13
8. Mill Square With Taper To Finish Dimension — 2" carb. end mill, 1 pass, 20 ipm. P6–P9
9. Mill Circle to Finish Dimension — 2" carb. end mill, 1 pass, 16 ipm. P2–P5

FIGURE 10-5
Detailed APT program (Courtesy of Cincinnati Milacron Inc.)

10. Mill Square Inside Circle to Finish Dimension — 2" carb. end mill, 1 pass, 16 ipm. P14–P17
11. Drill 4 Holes — 1-1/8" deep, 59/64 dia. drill. P18–P21
12. Pipe Tap 2 Holes — 3/4 — 14 pipe tap. P19 and P21
13. Bore 2 Holes — 1/2" deep, 1" dia. bore. P18 and P20
14. Drill 2 Holes — 7/8" deep, 5/16 dia. drill. P22 and P23
15. Tap 2 Holes — 3/8" deep, 3/8 — 16 n.c. tap. P22 and P23
16. Unload.

RAE-103

N/C 360 APT VERSION 4, MODIFICATION 2   DATE= 71.284 TIME OF DAY IN HRS./MIN./SEC IS 12/32/46.20

```
THE MACRO    CIR    USES    66  LOCATIONS IN CANON
THE MACRO    CIRSQR USES    90  LOCATIONS IN CANON
THE MACRO    ANGSQR USES    96  LOCATIONS IN CANON
THE MACRO    SQUR   USES    96  LOCATIONS IN CANON
```

TABLE USAGE DURING INPUT TRANSLATION

	PASS ONE		PASS TWO	
	ALLOCATED	USED	DYNAMIC ALLOCATION	
VST	2750	124	VST	125
PTPP	2225	272	PTPP	272
CANON	2225	348	SCALARS	11
			CANON	5792

```
 1  PARTNO   CPG COMPOSITE TEST PART FOR THE CIM-X 720 WITH TOOL COMP      CPGC0000
 2  $$                                                                    CPGC0010
 3           MACHIN/CINAC1,-15600C,CIRCUL,170,12,2                        CPGC0020
 4  $$                                                                    CPGC0030
 5           ORIGIN/0,0,3           $$ ORIGIN FOR CONTROL WITH TOOL COMP  CPGC0040
 6  $$                                                                    CPGC0050
 7  PT1    = POINT/0,0,0                                                  CPGC0060
 8  ULOAD  = POINT/15, 6,0                                                CPGC0070
 9  C1     = CIRCLE/0,0,5                                                 CPGC0080
10  HL0    = LINE/PT1,ATANGL,0.0                                         CPGC0090
11  VL0    = LINE/PT1,ATANGL,90.0                                        CPGC0100
12  AL0    = LINE/PT1,ATANGL,.0572222                                    CPGC0110
13  AL1    = LINE/PARLEL,AL0,YLARGE,5.25                                 CPGC0120
14  AL2    = LINE/PARLEL,AL0,YSMALL,5.25                                 CPGC0130
15  AL3    = LINE/PARLEL,(LINE/PT1,PERPTO,AL0),XLARGE,5.25               CPGC0140
16  AL4    = LINE/PARLEL,AL3,XSMALL,10.50                                CPGC0150
17  AL5    = LINE/POINT,YLARGE,INTOF,(ALA=LINE/PT1,ATANGL,55),C1),ATANGL,10 CPGC0160
18  AL6    = LINE/(POINT/YSMALL,INTOF,ALA,C1),ATANGL,10                  CPGC0170
19  ALB    = LINE/PT1,PERPTO,ALA                                         CPGC0180
20  AL7    = LINE/(POINT/XLARGE,INTOF,ALB,C1),PERPTO,AL5                 CPGC0190
21  AL8    = LINE/(POINT/XSMALL,INTOF,ALB,C1),PERPTO,AL5                 CPGC0200
22  VL1    = LINE/PARLEL,VL0,XLARGE,5.375                                CPGC0210
23  VL2    = LINE/PARLEL,VL0,XSMALL,5.375                                CPGC0220
24  VL3    = LINE/PARLEL,VL0,XLARGE,5.5                                  CPGC0230
25  VL4    = LINE/PARLEL,VL0,XSMALL,5.5                                  CPGC0240
26  HL1    = LINE/PARLEL,HL0,YSMALL,5.375                                CPGC0250
27  HL2    = LINE/PARLEL,HL0,YSMALL,5.5                                  CPGC0260
28  HL3    = LINE/PARLEL,HL0,YLARGE,5.375                                CPGC0270
29  HL4    = LINE/PARLEL,HL0,YLARGE,5.5                                  CPGC0280
30  PT18   = POINT/4.375,-4.375,-.625                                    CPGC0290
31  PT19   = POINT/4.375, 4.375,-.625                                    CPGC0300
32  PT20   = POINT/-4.375,4.375,-.625                                    CPGC0310
33  PT21   = POINT/-4.375,-4.375,-.625                                   CPGC0320
34  PT22   = POINT/2.5,-2.5,0.0                                          CPGC0330
35  PT23   = POINT/-2.5,2.5,0.0                                          CPGC0340
36  PT25   = POINT/3.1358,-7.0,-.3125                                    CPGC0350
37  PT26   = POINT/6.7600,-3.1358,-.3125                                 CPGC0360
38  PT27   = POINT/6.7600,3.1358,-.3125                                  CPGC0370
39  PT28   = POINT/3.1358,7.0,-.3125                                     CPGC0380
40  PT29   = POINT/-3.1358,7.0,-.3125                                    CPGC0390
41  PT30   = POINT/-6.7600,3.1358,-.3125                                 CPGC0400
42  PT31   = POINT/-6.7600,-3.1358,-.3125                                CPGC0410
```

VIII-D 1

FIGURE 10-5
Continued

```
 43   PT32    = POINT/-3.1358,-7.0,-.3125                               CPGC0420
 44   PT33    = POINT/3.1358,-7.0,0.0                                   CPGC0430
 45   L1      = LINE/PT1,PT25                                           CPGC0440
 46   PL0     = PLANE/0,0,1,0                                           CPGC0450
 47   PL1     = PLANE/0,0,1,-.3125                                      CPGC0460
 48   PL2     = PLANE/0,0,1,-.625                                       CPGC0470
 49   PL3     = PLANE/0,0,1,-.9375                                      CPGC0480
 50   PL4     = PLANE/0,0,1,-1.25                                       CPGC0490
 51   $$                                                                CPGC0500
 52   $$ THE FOLLOWING ARE SYMBOLIC FEEDRATES--CHECK TO SEE IF THEY SHOULD   CPGC0510
 53   $$ BE CHANGED PRIOR TO PROCESSING.                                CPGC0520
 54   $$                                                                CPGC0530
 55   RFED1   = 8                $$ RGH CORNERS OF PART, INSIDE SQUARE   CPGC0540
 56   RFED2   = 12               $$ RGT CIRCLE                          CPGC0550
 57   RFED3   = 14               $$ RGT OUTSIDE SQUARE                  CPGC0560
 58   FFED1   = 20               $$ FINISH MILL SQUARE                  CPGC0570
 59   FFED2   = 15               $$ FINSIH MILL CIRCLE                  CPGC0580
 60   FFED3   = 20               $$ FINISH TAPERD SQUARE                CPGC0590
 61   DRFED   = 15               $$ DRILL                               CPGC0600
 62   TPFED1  = 15               $$ TAPERD PIPE TAP                     CPGC0610
 63   TPFED2  = 21               $$ N.C.TAP                             CPGC0620
 64   BRFED   = 10               $$ BORE                                CPGC0630
 65   RAPD    = 150              $$ RAPID                               CPGC0640
 66   $$                                                                CPGC0650
 67   $$        MACROS                                                  CPGC0660
 68   $$                                                                CPGC0670
 69   CIR     = MACRO/DIA,PLN,FED                                       CPGC0680
 70           CUTTER/DIA                                                CPGC0690
 71           FEDRAT/FED                                                CPGC0700
 72           SPINDL/880,CLW                                            CPGC0710
 73           PSIS/PLN                                                  CPGC0720
 74           FROM/PT33                                                 CPGC0730
 75           INDIRP/PT1                                                CPGC0740
 76           GO/TO,C1                                                  CPGC0750
 77           TLRGT,GORGT/C1,ON,2,INTOF,(LINE/PT1,PT25)                 CPGC0760
 78           TERMAC                                                    CPGC0770
 79   $$                                                                CPGC0780
 80   CIRSQR  = MACRO/DIAM,FEDR                                         CPGC0790
 81           CUTTER/DIAM                                               CPGC0800
 82           CYCLE/MILLRP,DEEP,0.0                                     CPGC0810
 83           SPINDL/880,CLW                                            CPGC0820
 84           FEDRAT/FEDR                                               CPGC0830
 85           ZSURF/PL1                                                 CPGC0840
 86           GOTO/(POINT/INTOF,AL8,(LINE/PARLEL,AL7,XLARGE,1.25))      CPGC0850
 87           PSIS/PL1                                                  CPGC0860
 88           INDIRP/PT1                                                CPGC0870
 89           GU/TO,AL7                                                 CPGC0880
 90           TLRGT,GORGT/AL7                                           CPGC0890
 91           GOLFT/AL5                                                 CPGC0900
 92           GOLFT/AL8                                                 CPGC0910
 93           GOLFT/AL6,PAST,AL7                                        CPGC0920
 94           TERMAC                                                    CPGC0930
 95   $$                                                                CPGC0940
 96   ANGSQR  = MACRO/DIAMT,FDRT                                        CPGC0950
 97           CUTTER/DIAMT                                              CPGC0960
 98           SPINDL/880,CLW                                            CPGC0970
 99           FEDRAT/FDRT                                               CPGC0980
100           CYCLE/MILLRP,DEEP,0.0                                     CPGC0990
101           ZSURF/PL3                                                 CPGC1000
102           GOTO/(POINT/INTOF,(LINE/PARLEL,VL3,XLARGE,1.5),$         CPGC1010
                   (LINE/PARLEL,HL2,YSMALL,1.5))                        CPGC1020
103           PSIS/PL3                                                  CPGC1030
104           INDIRP/PT1                                                CPGC1040
105           GO/TO,AL3                                                 CPGC1050
106           TLRGT,GORGT/AL3                                           CPGC1060
107           GOLFT/AL1                                                 CPGC1070

108           GOLFT/AL4                                                 CPGC1080
109           GOLFT/AL2,PAST,AL3                                        CPGC1090
110           TERMAC                                                    CPGC1100
111   $$                                                                CPGC1110
112   SQUR    = MACRO/DTR,FRAT                                          CPGC1120
113           CUTTER/DTR                                                CPGC1130
114           SPINDL/880,CLW                                            CPGC1140
115           CYCLE/MILLRP,DEEP,0.0                                     CPGC1150
116           ZSURF/PL4                                                 CPGC1160
117           GOTO/(POINT/INTOF,(LINE/PARLEL,VL3,XLARGE,1.5),$         CPGC1170
                   (LINE/PARLEL,HL2,YSMALL,1.5))                        CPGC1180
118           FEDRAT/FRAT                                               CPGC1190
119           PSIS/PL4                                                  CPGC1200
120           INDIRP/PT1                                                CPGC1210
121           GO/TO,VL1                                                 CPGC1220
122           TLRGT,GORGT/VL1                                           CPGC1230
```

FIGURE 10-5
Continued

```
123            GOLFT/HL3                                              CPGC1240
124            GOLFT/VL2                                              CPGC1250
125            GOLFT/HL1,PAST,VL1                                     CPGC1260
126            TERMAC                                                 CPGC1270
127    $$                                                             CPGC1280
128    $$      START OF PROGRAM                                       CPGC1290
129    $$                                                             CPGC1300
130    PPRINT                                                         CPGC1310
131    PPRINT  ALIGN POINT IS AT 1INCH BORED HOLE (X000000 Y000000)   CPGC1320
132    PPRINT                                                         CPGC1330
133    PPRINT  LOAD 2-INCH DIA 4-FLUTE END MILL                       CPGC1340
134    PPRINT                                                         CPGC1350
135    PPRINT  MILL CORNERS OFF PART- 2 PASSES,5/16 DEPTH EACH PASS   CPGC1360
136            LOADTL / 1,1                                           CPGC1370
137    $$                                                             CPGC1380
138            CLRSRF / XYPLAN, 3                                     CPGC1390
139            ROTABL / ATANGL, 90                                    CPGC1400
140            SPINDL / 880, CLW                                      CPGC1410
141            COOLNT / MIST                                          CPGC1420
142            FEDRAT / RFED1                                         CPGC1430
143            FROM/PT1                                               CPGC1440
144    CYCLE/MILLRP,DEEP,0.0                                          CPGC1450
145            GOTO/PT25                                              CPGC1460
146            GOTO/PT26                                              CPGC1470
147            RAPID                                                  CPGC1480
148            GOTO/PT27                                              CPGC1490
149            GOTO/PT28                                              CPGC1500
150            RAPID                                                  CPGC1510
151            GOTO/PT29                                              CPGC1520
152            GOTO/PT30                                              CPGC1530
153            RAPID                                                  CPGC1540
154            GOTO/PT31                                              CPGC1550
155            GOTO/PT32                                              CPGC1560
156    CYCLE/MILLRP,DEEP,.3125                                        CPGC1570
157            GOTO/PT32                                              CPGC1580
158            GOTO/PT31                                              CPGC1590
159            RAPID                                                  CPGC1600
160            GOTO/PT30                                              CPGC1610
161            GOTO/PT29                                              CPGC1620
162            RAPID                                                  CPGC1630
163            GOTO/PT28                                              CPGC1640
164            GOTO/PT27                                              CPGC1650
165            RAPID                                                  CPGC1660
166            GOTO/PT26                                              CPGC1670
167            GOTO/PT25                                              CPGC1680
168    PPRINT                                                         CPGC1690
169    PPRINT  ROUGH MILL CIRCLE--2 DEPTH PASSES-- LEAVE .020 EXCESS--RPM 880  CPGC1700
170    PPRINT                                                         CPGC1710
171            CYCLE / MILLFD, DEEP, 0.3125                           CPGC1720
172            CALL/CIR,DIA=2.04,PLN=PL0,FED=RFED2                    CPGC1730
                                                                      CPGC1740
174            CYCLE/MILLFD,DEEP,.625                                 CPGC1750
175            CALL/CIR,DIA=2.04,PLN=PL0,FED=RFED2                    CPGC1760
176            GOTO/PT33                                              CPGC1770
177    PPRINT                                                         CPGC1780
178    PPRINT  MILL SQUARE INSIDE CIRCLE--ALLOW .020 EXCESS--RPM 880  CPGC1790
179    PPRINT                                                         CPGC1800
180            CALL/CIRSQR,DIAM=2.04,FEDR=RFED1                       CPGC1810
181    PPRINT                                                         CPGC1820
182    PPRINT  MILL TAPERD SQUARE--1 PASS-- LEAVE .020 EXCESS--RPM 880 CPGC1830
183    PPRINT                                                         CPGC1840
184            CALL/ANGSQR,DIAMT=2.04,FDRT=RFED3                      CPGC1850
185    PPRINT                                                         CPGC1860
186    PPRINT  MILL OUTSIDE SQUARE--1 PASS-- LEAVE .020 EXCESS--RPM 880 CPGC1870
187    PPRINT                                                         CPGC1880
188            CALL/SQUR,DTR=2.04,FRAT=RFED3                          CPGC1890
189    PPRINT                                                         CPGC1900
190    PPRINT  RETRACT TO BACK LIMIT--POSITION AT UNLOAD POINT        CPGC1910
191    PPRINT  INSPECT CUTTER AND PART                                CPGC1920
192    PPRINT                                                         CPGC1930
193            RAPID                                                  CPGC1940
194            GOTO/ULOAD                                             CPGC1950
195            OPSTOP                                                 CPGC1960
196    PPRINT                                                         CPGC1970
```

FIGURE 10-5
Continued

```
197   PPRINT MILL SQUARE--FINISH PASS--RPM 880          CPGC1980
198   PPRINT                                            CPGC1990
199         CALL/SQUR,DTR=2.0,FRAT=FFED3                CPGC2000
200   PPRINT                                            CPGC2010
201   PPRINT MILL TAPERD SQUARE -- FINISH PASS--RPM 880 CPGC2020
202   $$                                                CPGC2030
203         CALL/ANGSQR,DIAMT=2.0,FDRT=FFED3            CPGC2040
204   PPRINT                                            CPGC2050
205   PPRINT MILL CIRCLE--FINISH PASS--1 PASS--RPM 880  CPGC2060
206   $$                                                CPGC2070
207         CYCLE / MILLFD, DEEP, 0.625                 CPGC2080
208         GOTO/PT33,30                                CPGC2090
209         CALL/CIR,DIA=2.0,PLN=PLO,FED=FFED2          CPGC2100
210   PPRINT                                            CPGC2110
211   PPRINT MILL SQUARE INSCRIBED IN CIRCLE--FINISH PASS--RPM 880  CPGC2120
212   $$                                                CPGC2130
213         CALL/CTRSQR,DIAM=2.0,FEDR=FFED2             CPGC2140
214   PPRINT                                            CPGC2150
215   PPRINT LOAD TOOL 2 -- 59/64 DRILL                 CPGC2160
216   $$                                                CPGC2170
217         LOADTL / 2,2                                CPGC2180
218   $$                                                CPGC2190
219         SPINDL / 1400, CLW                          CPGC2200
220         COOLNT / FLOOD                              CPGC2210
221         CYCLE/DRILL,DEEP,1.1250,IPM,18.0            CPGC2220
222         GOTO/PT18                                   CPGC2230
223         RETRCT                                      CPGC2240
224         GOTO/PT21                                   CPGC2250
225         RETRCT                                      CPGC2260
226         GOTO/PT20                                   CPGC2270
227         RETRCT                                      CPGC2280
228         GOTO/PT19                                   CPGC2290
229   PPRINT                                            CPGC2300
230   PPRINT LOAD TOOL 3 -- 3/4-14 PIPE TAP             CPGC2310
231   $$                                                CPGC2320
232         LOADTL / 3,3                                CPGC2330
233         SELCTL / 4                                  CPGC2340
234         SPINDL / 175, CLW                           CPGC2350
235         CYCLE/TAP  ,DEEP,1.0,IPM,3.0                CPGC2360
236         GOTO/PT21                                   CPGC2370
237         RETRCT                                      CPGC2380
238         GOTO/PT19                                   CPGC2390
239   PPRINT                                            CPGC2400
240   PPRINT LOAD TOOL 4 -- 1.0 BORE(CARB. TIP)         CPGC2410
241   $$                                                CPGC2420
242         LOADTL / 4,4                                CPGC2430
243   $$                                                CPGC2440
244         SPINDL / 1755, CLW                          CPGC2450
245         CYCLE/BORE,DEEP,0.50,IPM,6.0                CPGC2460
246         GOTO/PT18                                   CPGC2470
247         RETRCT                                      CPGC2480
248         GOTO/PT20                                   CPGC2490
249   PPRINT                                            CPGC2500
250   PPRINT LOAD TOOL 5 -- 5/16 DRILL                  CPGC2510
251   $$                                                CPGC2520
252         LOADTL / 5,5                                CPGC2530
253   $$                                                CPGC2540
254         SPINDL / 2785, CLW                          CPGC2550
255         CYCLE/DRILL,DEEP,0.8750,IPM,12.0            CPGC2560
256         GOTO/PT22                                   CPGC2570
257         GOTO/PT23                                   CPGC2580
258   PPRINT                                            CPGC2590
259   PPRINT LOAD TOOL 6 -- 3/8-16 N.C. TAP             CPGC2600
260   $$                                                CPGC2610
261         LOADTL / 6,6                                CPGC2620
262         SPINDL / 220, CLW                           CPGC2630
263         CYCLE/TAP,DEEP,0.3750,IPM,3.0               CPGC2640
264         GOTO/PT22                                   CPGC2650
265         GOTO/PT23                                   CPGC2660
266         RAPID                                       CPGC2670
267         GOTO/ ULOAD                                 CPGC2680
268         END                                         CPGC2690
269         FINI                                        CPGC2700
```

FIGURE 10-5
Continued

71.284   C I N A C 1 156000 / A P T   I N C H   P O S T P R O C E S S O R   LEVEL A   PAGE   1

A C R A M A T I C 3 3 5 - D

O/N	G	X	Y	Z	I	J	F	R	S	B	W	M S	CLNO	RPM	TIME
CPG COMPOSITE TEST PART FOR THE CIM-X 720 WITH TOOL COMP															
LEADER/	72.0														

ALIGN POINT IS AT 1INCH BORED HOLE (X000000 Y000000)

LOAD 2-INCH DIA 4-FLUTE END MILL

MILL CORNERS OFF PART-- 2 PASSES, 5/16 DEPTH EACH PASS

O/N	G	X	Y	Z	I	J	F	R	S	B	W	M S	CLNO	RPM	TIME
$0 1	G80	X& 0	Y& 0	Z 0			F 1	R 0				M06$	32	110	.283
0 2	G80	X& 0	Y& 0	Z 0			F 80	R 30000	S 12	B 90000	W 1	M17$	32	880	.014
0 3	G80	X& 31358	Y- 70000	Z 0			F 80	R 34125	S 12	B 90000	W 1	M17$	36	880	.040
N 4	G79	X& 67600	Y- 31358									$	38	880	.662
N 5	G80		Y& 31358									$	42	880	.031
N 6	G79	X& 31358	Y& 70000									$	44	880	.662
N 7	G80	X- 31358										$	48	880	.031
N 8	G79	X- 67600	Y& 31358									$	50	880	.662
N 9	G80		Y- 31358									$	54	880	.031
N 10	G79	X- 31358	Y- 70000									$	56	880	.662
0 11	G80	X- 31358	Y- 70000	Z 0			F 80	R 37250	S 12	B 90000	W 1	M17$	60	880	.001
N 12	G79	X- 67600	Y- 31358									$	62	880	.662
N 13	G80		Y& 31358									$	66	880	.031
N 14	G79	X- 31358	Y& 70000									$	68	880	.662
N 15	G80	X& 31358										$	72	880	.031
N 16	G79	X& 67600	Y& 31358									$	74	880	.662
N 17	G80		Y& 31358									$	78	880	.031
N 18	G79	X& 31358	Y- 70000									$	80	880	.662

ROUGH MILL CIRCLE--2 DEPTH PASSES-- LEAVE .020 EXCESS--RPM 880

O/N	G	X	Y	Z	I	J	F	R	S	B	W	M S	CLNO	RPM	TIME
N 19	G80			Z 3125			F 120	R 30000				$	98	880	.003
0 20	G80	X& 31358	Y- 70000	Z 3125			F 120	R 30000	S 12	B 90000	W 1	M17$	98	880	.000
N 21	G79			Z 3125								$	98	880	.026
N 22		X& 24611	Y- 54939									$	101	880	.163
N 23	G03	X& 60200	Y& 0		I& 0	J& 0						$	107	880	.576
N 24	G03	X- 0	Y& 60200		I& 0	J& 0						$	107	880	.788
N 25	G03	X- 60200	Y- 0		I& 0	J& 0						$	107	880	.788
N 26	G03	X& 0	Y- 60200		I& 0	J& 0						$	107	880	.788
N 27	G03	X& 24611	Y- 54939		I& 0	J& 0						$	107	880	.211
N 28	G80	X& 31358	Y- 70000					R 33125				$	110	880	.009
N 29	G79			Z 3125								$	110	880	.026
N 30	G80											$	112	880	.000
N 31				Z 6250				R 30000				$	122	880	.001
0 32	G80	X& 31358	Y- 70000	Z 6250			F 120	R 30000	S 12	B 90000	W 1	M17$	122	880	.000
N 33	G79			Z 6250								$	122	880	.052
N 34		X& 24611	Y- 54939									$	125	880	.189
N 35	G03	X& 60200	Y& 0		I& 0	J& 0						$	131	880	.576
N 36	G03	X- 0	Y& 60200		I& 0	J& 0						$	131	880	.788

MACHINING TIME   10.815 MINUTES   TAPE LENGTH   6.70 FEET

71.284   C I N A C 1 156000 / A P T   I N C H   P O S T P R O C E S S O R   LEVEL A   PAGE   2

O/N	G	X	Y	Z	I	J	F	R	S	B	W	M S	CLNO	RPM	TIME
CPG COMPOSITE TEST PART FOR THE CIM-X 720 WITH TOOL COMP															
N 37	G03	X- 60200	Y- 0		I& 0	J& 0						$	131	880	.788
N 38	G03	X& 0	Y- 60200		I& 0	J& 0						$	131	880	.788
N 39	G03	X& 24611	Y- 54939		I& 0	J& 0						$	131	880	.211
N 40	G79	X& 31358	Y- 70000									$	134	880	.189

MILL SQUARE INSIDE CIRCLE--ALLOW .020 EXCESS--RPM 880

O/N	G	X	Y	Z	I	J	F	R	S	B	W	M S	CLNO	RPM	TIME
N 41	G80											$	145	880	.000
0 42	G80	X& 55438	Y- 38818	Z 0			F 80	R 34125	S 12	B 90000	W 1	M17$	151	880	.021
N 43	G79	X& 52774	Y- 36953									$	155	880	.040
N 44		X& 36953	Y& 52774									$	157	880	1.138
N 45		X- 52774	Y& 36953									$	159	880	1.138
N 46		X- 36953	Y- 52774									$	161	880	1.138
N 47		X& 52774	Y- 36953									$	163	880	1.138

MILL TAPERD SQUARE--1 PASS--LEAVE .020 EXCESS--RPM 880

O/N	G	X	Y	Z	I	J	F	R	S	B	W	M S	CLNO	RPM	TIME
0 48	G80	X& 70000	Y- 70000	Z 0			F 140	R 40375	S 12	B 90000	W 1	M17$	181	880	.021
N 49	G79	X& 62763	Y- 62763									$	185	880	.073
N 50		X& 62637	Y& 62763									$	187	880	.896
N 51		X- 62763	Y& 62637									$	189	880	.895
N 52		X- 62637	Y- 62763									$	191	880	.895
N 53		X& 62763	Y- 62637									$	193	880	.895

FIGURE 10-5
Continued

MILL OUTSIDE SQUARE--1 PASS-- LEAVE .020 EXCESS--RPM 880

```
0 54 G80 X&  70000 Y-  70000 Z     0              F 140 R  43500 S 12 B 90000 W 1 M17$ 209   880   .006
N 55 G79 X&  63950 Y-  63950                                                        $ 215   880   .061
N 56            Y&  63950                                                           $ 217   880   .913
N 57     X-  63950                                                                  $ 219   880   .913
N 58            Y-  63950                                                           $ 221   880   .913
N 59     X&  63950                                                                  $ 223   880   .913
```

RETRACT TO BACK LIMIT--POSITION AT UNLOAD POINT
INSPECT CUTTER AND PART

```
N 60 G80                                          R  31000                   $ 236   880   .006
N 61     X& 150000 Y& 60000                                                  $ 236   880   .075
N 62                                                                   M01$  238   880   .000
```

MACHINING TIME   24.893 MINUTES   TAPE LENGTH 11.12 FEET

MILL SQUARE--FINISH PASS--RPM 880

```
0 63 G80 X&  70000 Y-  70000 Z     0              F 140 R  43500 S 12 B 90000 W 1 M17$ 253   880   .082
N 64 G79 X&  63750 Y-  63077                       F 200                             $ 259   880   .046
N 65            Y&  63750                                                            $ 261   880   .634
N 66     X-  63750                                                                   $ 263   880   .637
N 67            Y-  63750                                                            $ 265   880   .637
N 68     X&  63750                                                                   $ 267   880   .637
```

MACHINING TIME   27.568 MINUTES   TAPE LENGTH 12.09 FEET

71.284   C I N A C 1 156000 / A P T   I N C H   P O S T P R O C E S S O R   LEVEL A   PAGE   3

CPG COMPOSITE TEST PART FOR THE CIM-X 720 WITH TOOL COMP

```
O/N G     X         Y         Z        I       J       F      R         S     B      W   M S CLNO   RPM    TIME
```

MILL TAPERD SQUARE -- FINISH PASS--RPM 880

```
N 69 G80                                          R  40375                   $ 283   880   .001
0 70 G80 X&  70000 Y-  70000 Z     0              F 200 R  40375 S 12 B 90000 W 1 M17$ 283   880   .004
N 71 G79 X&  62563 Y-  62694                                                 $ 287   880   .052
N 72     X&  62438 Y&  62562                                                 $ 289   880   .626
N 73     X-  62562 Y&  62438                                                 $ 291   880   .624
N 74     X-  62438 Y-  62562                                                 $ 293   880   .624
N 75     X&  62562 Y-  62438                                                 $ 295   880   .624
```

MILL CIRCLE--FINISH PASS--1 PASS--RPM 880

```
N 76 G80              Z  6250                      F 300 R  30000                   $ 305   880   .005
0 77 G80 X&  31358 Y-  70000 Z  6250              F 300 R  30000 S 12 B 90000 W 1 M17$ 305   880   .016
N 78 G79             Z- 6250                                                        $ 305   880   .020
0 79 G79 X&  31358 Y-  70000 Z  6250              F 160 R  30000 S 12 B 90000 W 1 M17$ 315   880   .039
N 80     X&  24532 Y- 54761                                                         $ 318   880   .143
N 81 G03 X&  60005 Y&     0    I&   0 J&   0                                        $ 324   880   .431
N 82 G03 X-      0 Y&  60005  I&   0 J&   0                                         $ 324   880   .589
N 83 G03 X-  60005 Y-     0    I&   0 J&   0                                        $ 324   880   .589
N 84 G03     X&      0 Y-  60005  I&   0 J&   0                                     $ 324   880   .589
N 85 G03 X&  24532 Y- 54761   I&   0 J&   0                                         $ 324   880   .157
```

MILL SQUARE INSCRIBED IN CIRCLE--FINISH PASS--RPM 880

```
N 86 G80                                                                     $ 334   880   .000
0 87 G80 X&  55438 Y-  38818 Z     0              F 160 R  34125 S 12 B 90000 W 1 M17$ 340   880   .019
N 88 G79 X&  52544 Y-  36801                                                 $ 346   880   .022
N 89     X&  36790 Y&  52542                                                 $ 346   880   .567
N 90     X-  52542 Y&  36790                                                 $ 348   880   .566
N 91     X-  36790 Y-  52542                                                 $ 350   880   .566
N 92     X&  52542 Y-  36790                                                 $ 352   880   .566
```

LOAD TOOL 2 -- 59/64 DRILL

```
0 93 G80 X&  52542 Y-  36790 Z     0              F 160 R     0          W 1 M06$ 367   880   .117
0 94 G81 X&  43750 Y-  43750 Z -11250             F 180 R  36250 S 14 B 90000 W 2 M13$ 367  1400   .091
N 95 G80                                          R  36250                   $ 369  1400   .018
N 96 G81 X-  43750                                R  36250                   $ 371  1400   .129
N 97 G80                                          R  36250                   $ 373  1400   .018
N 98 G81            Y& 43750                       R  36250                   $ 375  1400   .129
N 99 G80                                          R  36250                   $ 377  1400   .018
N100 G81 X&  43750                                R  36250                   $ 379  1400   .129
```

LOAD TOOL 3 -- 3/4-14 PIPE TAP

```
0101 G80 X&  43750 Y&  43750 Z     0              F 180 R     0          W 2 M06$ 393  1400   .118
0102 G84 X-  43750 Y-  43750 Z -10000             F  30 R  36250 S  3 B 90000 W 3 M13$ 393   175   .746
N103 G80                                          R     0                    $ 395   175   .018
N104 G84 X&  43750 Y&  43750 Z                    R  36250                   $ 397   175   .746
```

LOAD TOOL 4 -- 1.0 BORE (CARB. TIP)

```
0105 G80 X&  43750 Y&  43750 Z     0              F  30 R     0          W 3 M06$ 409   175   .118
0106 G85 X&  43750 Y-  43750 Z -5000              F  60 R  36250 S 15 B 90000 W 4 M13$ 409  1755   .228
N107 G80                                          R     0                    $ 411  1755   .018
```

MACHINING TIME   37.665 MINUTES   TAPE LENGTH 20.10 FEET

FIGURE 10-5
Continued

```
71.284            C I N A C 1  156000 / A P T     I N C H     P O S T P R O C E S S O R   LEVEL  A           PAGE    4

  CPG COMPOSITE TEST PART FOR THE CIM-X 720 WITH TOOL COMP
  O/N  G      X         Y          Z       I      J        F        R      S     B     W   M  S CLNO   RPM    TIME
  N108 G85 X-  43750 Y&  43750                              R  36250                       $  413  1755   .246

  LOAD TOOL 5 -- 5/16 DRILL
  0109 G80 X-  43750 Y&  43750 Z     0                  F  60 R     0                   W 4 M06$  425  1755   .118
  0110 G81 X&  25000 Y-  25000 Z   8750                 F 120 R 30000 S 17 B 90000 W 5 M13$  425  2785   .140
  N111     X-  25000 Y&  25000                                                          $  427  2785   .112

  LOAD TOOL 6 -- 3/8-16 N.C. TAP
  0112 G80 X-  25000 Y&  25000 Z     0                  F 120 R     0                   W 5 M06$  439  2785   .114
  0113 G84 X&  25000 Y-  25000 Z  -3750                 F  30 R 30000 S  4 B 90000 W 6 M13$  439   220   .300
  N114     X-  25000 Y&  25000                                                          $  441   220   .285
  N115 G80 X& 150000 Y& 60000                                                           $  445   220   .089
  N116                                                                            M02$  447   220   .000
  LEADER/   72.0
  CPG COMPOSITE TEST PART FOR THE CIM-X 720 WITH TOOL COMP

  MACHINING TIME   39.074 MINUTES                          TAPE LENGTH   22.32 FEET
```

** END OF POST PROCESSING **

FIGURE 10-5
Continued

ELAPSED TIME IS   0.36300 MINUTES

and understanding of computers and numerical control can be gained once these terms are defined and understood.

Since the beginning of N/C, approximately 1954 until the early 1970s, all MCUs were *hard-wired.* This meant that all of the logic was built-in and determined by the physical electronic elements of the control unit. These elements then controlled all functions such as tape format, absolute or incremental positioning, and character code recognition. In the early 1970s, more capable and less expensive electronics began to emerge. These types of computer elements, or complete minicomputers, became part of the control units. Functions that were solely the result of hardware design became resident in computer elements within the control unit.

The totality of soft-wired controls became effective at the 1976 machine tool show. What had once been the result of hardware design was replaced with complete computer logic, which had more capability, was no more costly, and could be programmed for a variety of functions at any time. Essentially, the ability of a machine control unit to recognize different tape formats was not locked in at the time of manufacture. It was simply a matter of how the computer element within the control unit was programmed to read the various tape codes, functions, etc.

The physical components for the soft-wired CNC units are the same whether they are controlling a lathe or a machining center. It is not the control unit elements, but rather the *executive program* or load tape that makes a control unit think like a machining center or lathe. Thus, the basic functioning of the soft-wired or CNC unit can be altered by changing the executive or load program. This executive program is supplied by the control unit manufacturers. In most cases, the user does not attempt to alter this executive program in any way.

The executive program can be revised, updated, or modified at any time. New operations can be introduced by simply reading them into the executive portion of the control unit computer. The input medium for this process is usually punched tape. Changing functions in a hard-wired control would involve changing the particular controlling elements within the control unit structure.

CNC VERSUS DNC

There is no greater evolutionary aspect of numerical control than that of the control unit. Control units have progressed from bulky tube types in the early 1950s to the microprocessor-based CNC units of today.

Actually, the CNC units evolved from the direct numerical control (DNC) applications of the late 1960s and early 1970s. The early DNC systems were capable of controlling a large number of machine tools. They provided some form of tape editing at the machine tool site, and they collectively controlled production and machine tool status. With a DNC system, the tape reader is bypassed. If the "host" computer went down, all machines could still be operated independently via tape input. An example of a typical DNC setup is illustrated in figure 10-6.

Some companies were reluctant to accept DNC because it really did not do much more than replace the tape reader. Even then, it was used only when the computer was running. Another problem in the early days of DNC was that computers capable of handling a DNC system were relatively expensive and not totally reliable. The idea of tying up twenty or more machines to a

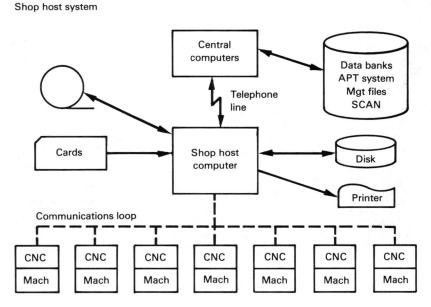

FIGURE 10-6
Typical DNC setup (host computer)

host computer and having them all go idle discouraged many prospective DNC customers. Not long after the initial DNC offerings, the minicomputer was developed. It provided some form of relief to both the cost and reliability problems of DNC.

As DNC became somewhat sidetracked and with the advent of more capable and less expensive minicomputers, computer technology advanced rapidly. Control builders quickly capitalized on these developments by incorporating minicomputers, and then microcomputers, into intelligent controls that have come to be known as CNC.

Both current and future developments for DNC revolve around computer software. Computer software traditionally lags computer hardware by several years. Computer technology has accelerated so rapidly that software has fallen even further behind. It will undoubtedly take several years for factory management systems to be developed from a software point of view that take full advantage of increased computer capability. A further discussion of DNC and broader computer applications will be discussed in Chapter 12.

A   B

FIGURE 10-7
Typical CNC units (A, Courtesy of Industrial Controls Division, Bendix Corporation; B, Courtesy of Cincinnati Milacron Inc.)

CNC, as discussed, was an offshoot of DNC. CNC has literally taken some of the abilities of a computer and, in a very compact package, applied them to dedicated service of a single N/C machine.

Many CNCs look the same, however, a minicomputer is built into each of these controls. Minicomputers may have 8,000 to 32,000 word capacity; they may contain even more. This capacity is used by advanced integrated circuitry made from chips like those used in hand calculators. Examples of typical CNC units are shown in figure 10-7.

The CNC unit of today has several characteristics not found in the traditional hard-wired control. Figure 10-8a illustrates how a CNC functions through its minicomputer. Figure 10-8b shows a typical CNC with machine tool.

Some of the important features of a CNC unit are as follows:

Memory. It is possible to "read in" a length of workpiece program tape. The amount of program tape depends on computer capacity, but the average length is around one hundred to two hundred feet. Once the program is in computer memory, it is possible to run each workpiece from the stored data, thus eliminating the usual requirement of reading the tape. In addition, programs may be added manually (MDI) and executed from memory as well.

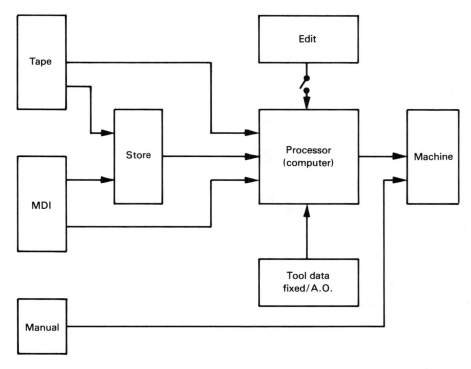

FIGURE 10-8a
How a CNC unit functions through its minicomputer

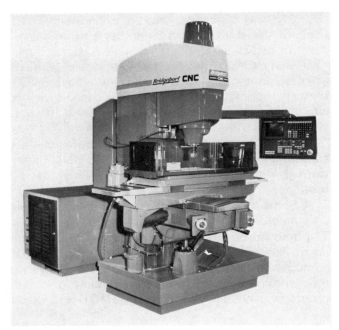

FIGURE 10-8b
Typical CNC with machine tool
(Courtesy of Bridgeport
Machines)

Edit. This feature adds the capability of "overriding" the tape if the tape input is being used. It also makes changes in the tape if the tape has been stored in computer memory.

CRT (Cathode Ray Tube). This is similar to a small television screen. Words and numbers appear on the screen, which displays pertinent information about the program. There is also a keyboard which enables communication with the control unit. A typical CRT screen, the type of information displayed, and an example keyboard is shown in figure 10-9.

Diagnostics. This capability refers to troubleshooting features as part of the control unit. With recent advances in *diagnostic tests,* if the CNC unit "goes down," it is only a short time before the problem is identified and corrected.

Two methods of diagnosing the difficulties are used. One method uses a special diagnostic tape which is supplied with the CNC unit. This tape checks many different elements and, on the CRT (or a separate oscilloscope) or by means of signal lights, indicates when the trouble occurred and where it is. However, the diagnostic tape is limited in its ability to search and signal. The second method is the added feature of being connected by telephone to the CNC manufacturer. They are able to run a variety of tests and can usually spot the source of the problem.

Other features available on CNC units are:

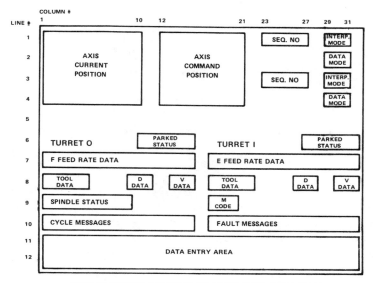

DISPLAY FORMAT FOR FULLY EXPANDED FOUR AXIS SYSTEM

CRT Display Format

Keyboard

FIGURE 10-9
Typical CRT screen with information displayed and keyboard

Tool Gaging Systems. These interface tool data through electronic tool gages directly to the CNC unit. The computing power of the CNC unit can be used to improve directly the potential productivity of the machine tool. This can be accomplished by tool identification and gaging, tool data entry, and tool matrix loading.

Corrected Tape. Optional tape punches exist for most CNC units so that once edited information proves correct for part production, a new tape can be punched at the machine from computer memory.

Special Routines. Some CNC executive programs are written so they will compute common routines such as bolt hole circles and pocketing routines from a single descriptive statement.

Incremental or Absolute. CNC units have the capacity to handle either type of command upon a tape code designation.

EIA versus ASCII. Many new CNC machine units will read either the EIA or ASCII code standard, and identify which one it is through parity check. It is not necessary to set selector switches or indicate the type of code.

Inch/Metric. Most CNC units and many of the hard-wired units have both inch and metric controlling capabilities. They may be a switch or a specific instruction within the workpiece program.

Future. It is conceivable that, in the near future, entire processor languages such as APT could be resident in the control units. They could then operate entirely from the program statements rather than using a remote computer to generate specific machine commands. In addition, computer graphics capabilities are currently being tested for visual display and manipulation of part geometry through the CNCs own CRT. Some of these advanced CNC units have made their way into the marketplace, and growth is expected to continue in the future.

REVIEW QUESTIONS

1. How is the computer used in numerical control applications?
2. Name some N/C processor languages available and explain the differences between them.
3. How did the APT language evolve? What steps are being taken to improve its effectiveness?
4. What are the four major sections of the APT system? Briefly explain their functions.
5. What is a postprocessor? How does it relate to the overall APT system?
6. What are the major functions of a postprocessor?
7. What types of statements are used when programming in the APT language? Explain.
8. Explain the differences between hard-wired and soft-wired numerical controls.
9. What is the difference between DNC and CNC?
10. Describe some of the features available on CNC units. Briefly explain their functions.

CHAPTER 11

Tooling for Numerical Control Machines

OBJECTIVES   After studying this chapter, the student will be able to:

- Understand the overall importance and impact of proper tooling on an N/C machine.
- Explain how correct use of cutting tools affects overall machine performance and productivity levels.
- Discuss how proper fixturing leads to successful use of an N/C machine.
- Identify sound tooling practices for productive part processing capabilities.

TOOLING CONSIDERATIONS

Tooling for N/C machines has always been one of the most neglected elements of an N/C installation. When planning and justifying N/C equipment, this aspect of tooling is often given secondary consideration. This is because all tooling tends to be taken for granted until something goes wrong.

N/C machines can only move or position appropriate cutting tools to specific locations, and rotate them or the workpiece at desired spindle speeds. The individual cutting tools actually do the metal removal work. The only way an N/C machine can be efficiently and effectively used is through proper use and care of cutting tools and work-holding devices.

In conventional machining, part accuracies depend on special fixturing. This type of fixturing has precisely made, precisely located tool-setting pads, and accurately located bushings that guide the tools. With an N/C machining center, simple fixturing is used. There are no tool bushings or tool pads to guide the tools. The repetitive positioning accuracy of the machine promises a high degree of quality. However, machining accuracies depend on the inherent accuracies of the cutting tools and their holders. If a drill "runs out," the benefit of the machining center's accuracy is lost. The programmer must assume that the tools will not run out.

There is another reason for careful selection of cutting tools. The average, conventional machine tool cuts metal only 20% of the time. An N/C machining center can be expected to cut metal up to 75% of the time. This results in more tool usage in a given period of time. Tool life, measured in "time in the cut," will be as good or better but, because of the increased usage, cutting tools will be used up three times as fast. The cost of perishable tools used during the machine's lifetime may amount to 50% or more of the machine's purchase price. Therefore, perishable tools represent a sizable investment; hence the importance of getting high volume along with good tool life.

It should be noted that an N/C machine is no more accurate than the cutting tools used with it to machine the workpiece. Thus, the decision for buying cutting tools and tool holders should receive the same consideration as was given the purchase of the machine.

CUTTING TOOLS USED ON N/C EQUIPMENT

A variety of cutting tools are used on N/C equipment to perform a multitude of machining operations. Many of the cutting tool applications, however, are no different than that which would have to be performed on manual equipment to produce the same workpiece. Cutting tools range from conventional drills, taps, end mills, to high-technology carbide cutting tools. Because of the importance of cutting tools to the overall manufacturing process and their costs, it is important that each be examined in detail.

DRILLS

Even though the slide positioning accuracy of most modern machining centers is ±.001 or better, there is no guarantee that the drilled hole location will be within that degree of accuracy. A standard, commercial twist drill, manufactured to specifications, may be very accurate, or it may be so inaccurate that nothing more than roughing work is possible.

All new drills have certain allowable tolerances, as depicted in figure 11-1. Those tolerances that affect accuracy the most are lip height, web

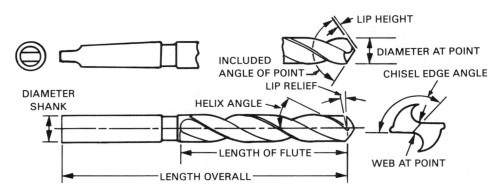

FIGURE 11-1
Identity of toleranced areas for two-flute general purpose drills

centrality, and flute spacing. The lip height, for example, of a .250-inch drill can vary 0.004 inch, its web can be off center as much as 0.005 inch, its flute spacing can be off by 0.006 inch, and the drill will still be within required specifications. Since a .250-inch drill is normally fed at a rate of .004 or .005 IPR, it would be impossible for that drill to produce accurate holes.

Since approximately 70% of all hole making is drilling, tool selection is of primary importance. One of the most important criterion in selecting a drill is to choose the shortest drill length that will permit drilling the hole to the desired depth. A good rule to remember is: the smaller the drill size, the smaller the allowable error; as drill size increases, the allowable error progressively increases. Short, stubby drills run truer, allow the fastest feeds, and improve tool life. The torsional rigidity of a drill will affect not only tool life and feed potential, but hole quality as well. Torsional rigidity is a measure of the tool's ability to resist twisting or unwinding; rigidity increases as drill length decreases. Therefore, on machining centers, where feed is constant and rapid, a shortened flute becomes a distinct advantage.

Many different types and varieties of drills exist and are used for a wide variety of applications. The common twist drill certainly has its applications but so do center drills, spade drills, and subland drills. Center drills are primarily designed to produce accurate centers in the work so that follow-up drills will start in perfect alignment. The proper selection and use of these drills will increase the accuracy of hole location, particularly on rough surfaces.

Ideally, the center drilled hole should be machined to a depth where the countersunk portion is 0.003 to 0.006 inches larger than the finished hole size, figure 11-2. With this method, the drill periphery will be guided into the countersunk hole, the location will be accurate, and the finished hole will have a chamfered or deburred edge.

The most widely used center drill is the bell or combination type, figure 11-3. It is commonly used in lathe work to provide work centers for subsequent operations. The advantages of this drill are accuracy and availability. However, for work assigned to an N/C machine, it has two disadvantages. First, the lead portion of the drill breaks off quite easily. Second, the drill is limited to small diameters.

If large-diameter holes are required to be machined to relatively accurate tolerances, the twist drill may be impractical. Spade drills are sometimes considered, as illustrated in figure 11-4, because they can produce large holes in one pass. In contrast, there is the conventional twist drill which makes progressively larger holes until the desired size is obtained.

Spade drills are advantageous in N/C work because only the blade, not the entire tool, needs to be changed

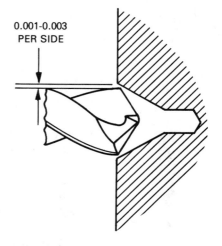

0.001-0.003 PER SIDE

FIGURE 11-2
Ideal center drill size in relation to finish hole size

FIGURE 11-3
**Bell or combination center drill (Courtesy of
Cleveland Twist Drill)**

FIGURE 11-4
**Spade drill showing blade and shank (Cour-
tesy of DoAll Company)**

FIGURE 11-5
**A multiple-diameter
(subland) drill
(Courtesy of Cleve-
land Twist Drill)**

when it becomes dull. Thus, correct tool length is maintained, and reset-
ting the tool length or recompensating the machine is eliminated.

The spade drill will normally use the same feeds and speeds as a twist
drill. In cast iron, the spade drill performs well at almost any depth.
However, in steel and aluminum, if the hole depth is more than one and
one-half or two times the hole diameter, problems with heat and chip
removal can occur.

Price becomes another important consideration in drill purchase and
selection. Generally, on a range of drill sizes from one to two inches (by
32nds), twist drills cost twice as much as spade drills. This is mainly because various blade sizes are interchangeable in a single shank. For example, only three spade drill shanks are required to hold the entire one- to
two-inch-blade range.

Many hole-producing sequences require multiple operations on the
same hole, such as drill and countersink, drill and counterbore, or drill and
body drill. Multiple-diameter, multiple-land tools, called subland tools,
are commonly used today, figure 11-5. The proper use of this type of
drill can result in a time savings and quality improvement. By combining
multiple-drilling operations into one tool, extra machining time and tool
handling time are eliminated. An additional benefit is derived from the
rigidity of the larger diameter, both in the ability to use maximum feed
rates and in improved hole accuracy.

TAPS

Tapping is one of the most difficult machining operations because of the ever present problem of chip clearance and adequate lubrication at the cutting edge of the tap. This is further aggravated by coarse threads in small diameters, long-thread engagements, unnecessarily high-thread percentages, tough materials, and countless other factors. Further, the relationship between speed and feed is fixed by the lead of the tap and cannot be varied independently. Tapping on an N/C machine frees the operator of the skill needed to tap a good hole since the tapping operation is programmed. Therefore, the prime concern is the tap, not the skill.

Generally, taps are divided into two major classifications: hand taps and machine screw taps. Their names, however, do not denote the manner in which the taps are used, because they are both used in power tapping of drilled holes.

Hand taps, figure 11-6, were originally intended for hand operation, but now they are widely used in machine production work. The name denotes the group of taps that are available in fractional sizes. The most commonly used hand taps include sizes ranging from 1/4 to 1 1/2 inches.

Machine screw taps is the name given to the group of taps available in decimal sizes. Machine screw taps are actually small hand taps. Their size is indicated by the machine screw system of sizes, ranging from #0 to #20. In this system, #0 is equivalent to 0.060 inch with a regular incremental increase of 0.013 inch between sizes. Therefore, #1 equals 0.073 inch, #2 equals 0.086 inch, and so on. The most commonly used machine screw taps are those between *numbers* #0 and #14, excluding #7 and #9.

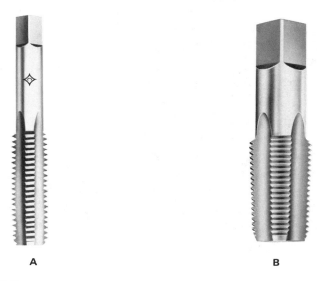

A   B

FIGURE 11-6
Conventional hand taps (A, Courtesy of Cleveland Twist Drill; B, Courtesy of Morse Cutting Tools Division, Gulf & Western Manufacturing Company)

FIGURE 11-7
Spiral pointed, gun, or chip-driver tap (Courtesy of DoAll Company)

Spiral pointed taps, figure 11-7, are sometimes referred to as gun, chip driver, or cam point taps. They are recommended for through holes and for holes with sufficient clearance at the bottom to provide chip space. Spiral pointed taps have straight flutes with a secondary grind in the flutes along the chamfer. This is ground into the flute at an angle to the axis of the tap so that it produces a shearing action when cutting the thread. As a result of this shearing action, the chips are forced ahead of the tap with very little resistance to thrust. The shearing action allows additional strength to be designed into the tap by making the web heavier and the flutes smaller than on conventional taps.

The main advantage of the spiral pointed tap is that it prevents chips from packing in the flutes or wedging between the flanks and the work. This is a major cause of tap breakage, particularly in small taps. The spiral pointed tap also allows a better flow of lubricant to the cutting edges. Tests have proven that spiral pointed taps require much less power than conventional straight-fluted taps, thus further reducing possible tap breakage.

Spiral-fluted taps, figure 11-8, are recommended for tapping blind holes where the problem of chip elimination is critical. They are most effective when the material being tapped produces long, stringy, curling chips. The spiral-fluted tap cuts freely while ejecting chips from the tapped hole. This prevents clogging and damage to both the threads and the tap. Chip removal is accomplished by the backward thrust action of the spiral flutes.

These taps are designed in much the same manner as conventional twist drills. They have helical flutes that lift the chips out of the hole.

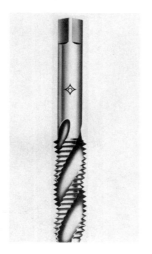

A

B

FIGURE 11-8
Spiral-fluted or turbo taps (A, Courtesy of Cleveland Twist Drill; B, Courtesy of DoAll Company)

Although it is incorrect to identify helical flutes as spiral flutes, both terms indicate helical flutes and are used interchangeably.

Spiral-fluted plug taps generally offer better results than straight-fluted taps. Three to five threads of the tap are chamfered. If trouble is encountered when tapping soft, stringy materials, a spiral-fluted bottoming tap is used. This tap has only one to one and one-half threads chamfer; thus, it produces heavier chips. The heavier chips are lifted more easily out of the hole than are the greater number of lighter chips produced by the plug tap.

In tapping operations, there are many factors which reduce tap life. Studies of the relating factors indicate that most problems stem from one area: eliminating chips from the hole. However, with the relatively new fluteless taps, figure 11-9, this problem and others can be solved. Fluteless taps do not produce chips. Rather, they form or roll the threads into the hole through cammed lobes on the periphery of the tap. Because of this forming action, the tap drill used is always larger than that used for a conventional tap. Fluteless tapping requires one deviation from normal tapping practice. Since it is basically a forming rather than a cutting operation, a high-pressure lubricant should be used rather than a cutting compound. This is extremely important for tapping the tougher materials when fluteless taps are used.

Traditionally certain materials have been tapped dry, plastics and cast iron for example. Even where dry tapping is possible, improved performance is generally affected by a judicious selection of some type of lubricant. Logically, lubrication is one of the key elements in successful tapping. It must therefore receive more than casual attention.

Tapping lubricants serve several purposes, the most important being:

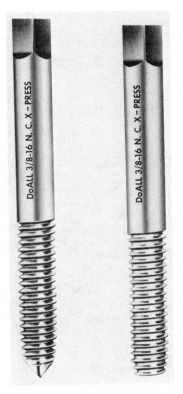

FIGURE 11-9
Fluteless taps (Courtesy of DoAll Company)

- They reduce friction.
- They produce clean, accurate threads by washing chips out of the tap flutes and the threaded hole.
- They improve the thread's surface finish.
- They reduce build-up on edges or chip welding on the cutting portion of the tap.

In summary, when purchasing a tap, you are, in effect, buying tapped holes rather than taps. Thus, tap life becomes the primary concern.

REAMING

Reaming is the process of removing a small amount of material, usually .062 inch or less, from a previously produced hole. The reamer is a

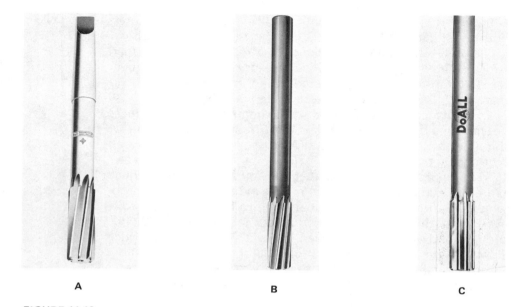

A   B   C

FIGURE 11-10
Some typical reamers — spiral and straight flutes (A, Courtesy of Cleveland Twist Drill; B, Courtesy of Morse Cutting Tools Division, Gulf & Western Manufacturing Company; C, Courtesy of DoAll Company)

multibladed cutting tool designed to enlarge and finish a hole to an exact size. Since the reamer is basically an end cutting tool mounted on a flexible shank, it cannot correct errors in hole location, hole crookedness, etc. A reamer will follow a previously produced hole. Therefore, when straightness or location is critical, some prior operation, such as boring, must have been performed in order to obtain these qualities.

Reamers can have either straight or spiral flutes and either a right- or left-hand helix, figure 11-10. Those with spiral or helical flutes will ordinarily provide smoother shear cutting and a better finish.

The shell reamer, figure 11-11, is primarily used for sizing and finishing operations on large holes, usually 3/4 inch and larger. The reamer will fit either a straight or taper shank arbor. Several different sizes of shell reamers may be fitted to the same arbor. This results in a tool savings.

The rose chucking and standard chucking reamers are one-piece construction and appear identical. Straight-fluted chucking reamers are used for free-cutting materials and finishing operations. Rose chucking reamers are similar to the standard chucking reamers. However, they are used primarily for roughing operations and operations where heavy stock removal is desired.

Single-blade floating reamers are becoming more popular for hole-finishing operations, particularly large holes. They also save tool inventory; with only eleven shanks or arbors and separate blades, all hole sizes from 5/8 inch to approximately six inches can be reamed.

Reaming on an N/C machine is an accepted practice. Floating tool-holders usually are not required for reaming operations on an N/C machining

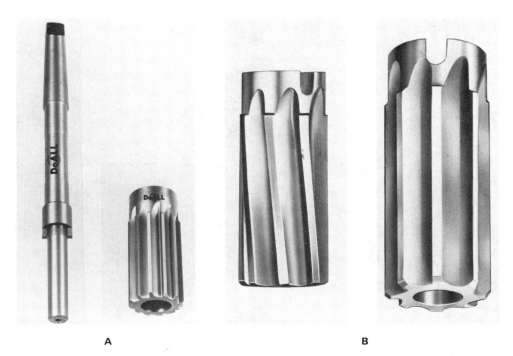

FIGURE 11-11
Shell reamers (A, Courtesy of DoAll Company; B, Courtesy of Morse Cutting Tools Division, Gulf &
Western Manufacturing Company)

center for two reasons. First, the repetitive positioning accuracy of the
machine slides ensures that the reamer will be positioned in line with the
hole. Second, the accuracy of present day collets and adaptors ensures that
the reamer will run "true," and therefore cut to size.

BORING

Boring is a machining operation that implies extreme accuracies. Therefore, a boring machine should be capable of positioning within tenths and
capable of holding roundness within millionths. If roundness within two or
three tenths and positioning accuracy within ±.0005 or ±.001 is satisfactory,
then some N/C machining centers can be used as a boring machine. Even
then, these tolerances are very difficult to maintain without a thorough
knowledge of boring methods and boring tools.

Boring operations are performed to produce accurate diameters, accurate
locations, good finishes, and true, straight holes. Properly performed, boring
is the one hole-finishing process whereby the full positioning accuracy of an
N/C machine can be used. A cored hole may be cast out of location or, in
drilling, the drill may wander beyond the acceptable tolerance. Boring can
correct these errors and finish the hole with a high degree of accuracy.

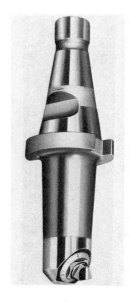

A   B

FIGURE 11-12
Typical boring bars with cartridge inserts (A, Courtesy of Cincinnati Milacron Inc.; B, Courtesy of Kennametal, Inc.)

A variety of boring bars similar to the ones in figure 11-12 and numerous types of cutters are used for modern boring operations. Regardless of the type used, all have certain common characteristics. Every boring cutter has a side cutting edge and an end cutting edge. These edges are related to the tool shank and are part of the standard nomenclature of single-point cutting tools of the American National Standards Institute (ANSI).

The type of boring to be done determines to an extent, the boring tools needed. For example, of the average parts processed on a machining center, 70% are drilling parts, 20% are milling parts, and 10% are boring parts. If boring is only an occasional operation, an offset type boring head may be used. Even though it has limited stock-removal capability, the wide adjustment range is a definite advantage. If boring requirements are on more of a production basis and cost must be kept to a minimum, then cartridge-type cutters are more economical. Turning centers also use a wide variety of boring bars and inserts. Consult local tooling vendors for additional, up-to-date boring bar and cutter information.

Another element of boring which deserves consideration is the length-to-diameter ratio (L/D). This ratio is one of the most important, but most neglected and least understood, aspects of boring. It refers to the length of the boring bar in relation to its diameter. Some of the largest manufacturers of boring machines and boring bars have researched this problem extensively. Studies indicate that a boring bar with a 1:1 length-to-diameter ratio is 64 times more rigid than a boring bar with a 4:1 ratio; it is 343 times more rigid than one with a 7:1 ratio. It follows, then, that in order to obtain maximum rigidity and accuracy from boring operations, the boring bar should be as short as possible.

Regardless of what is done to make the machine tool rigid enough for boring, if the same consideration is not applied to the tooling used inaccuracies will result. A standard heat-treated tool steel boring bar has a modulus of elasticity of about 30,000,000 pounds per square inch. Carbide, for instance, has a modulus of elasticity of 94,000,000 pounds per square inch. Thus, the boring bar made of solid carbide is over three times as rigid as a steel boring bar.

MILLING

With the exception of drills, probably the most widely used and efficient metal-removal tool for a machining center is the end mill. While arbor-mounted milling cutters on long production runs produce cheaper chips per dollar's worth of tool, the end mill is usually more economical for job shop quantities. Since machining centers are most efficient for short- to medium-sized production runs, it is evident that an end mill similar to that in figure 11-13 takes its place as one of the basic tools for machining centers.

Machining centers, by their very design, are capable of bringing the cutting tool to the workpiece with more accuracy, horsepower, and rigidity than conventional machines. Consequently, for milling operations, the limiting factors would be the setup, the toolholder, and the cutter rather than the machine itself.

Considering the work potential of end mills, a machining center's contouring capability should be used often. Many parts have milled surfaces bored holes, recesses, cobores, face grooves, and pockets; all of these operations are relatively easy to perform with two- or three-axis contouring and circular interpolation available on most modern machining centers.

When milling a flat face, contouring can help by allowing optimum cutter paths and not restricting milling to straight-line cuts. If a flat face with irregular edges must be machined, the cutter path can be programmed to follow the edges with no loss in efficiency and possibly a decrease in actual cutting time. In many cases, bored holes can be rough and semifinish bored by programming an appropriate end mill in a circular path around the centerline of the bore. This works well where the depth of the bored hole is about 1/3 of the cutter diameter or less. For bores of greater depth, a cut of 3/8- to 1/2-inch deep should be programmed, returned to center, moved in another 3/8 to 1/2 inch, and the circular cut repeated. This process can then

FIGURE 11-13
Double-end end mill (Courtesy of Sharpaloy Division, Precision Industries, Inc., Centerdale, R.I.)

be repeated several times, consistent with the flute length of the end mill. A practical limitation would be for holes where the depth does not exceed about 2/3 of the cutter diameter. This method can result in some tangible savings. One end mill can replace two or more boring bars, and one end mill can be used for several different bores. Tool drum storage space can be freed for additional tools, tool inventory reduced, and some time can be saved.

Many other types of end mills may be used on N/C machines, such as shell end mills with serrated and indexable blades, and face mills for a variety of applications. However, the success or failure of any milling operation depends largely on cutter life. When the cutter fails to produce an acceptable part, the milling machine or machining center must be shut down until the cutter is reconditioned or replaced.

COUNTERSINKING AND COUNTERBORING

Countersinking on N/C machines can be a frustrating experience because of the difficulty in establishing set lengths. To set up a job accurately using a standard, single-flute, nonpiloted countersink, figure 11-14, optical measuring equipment must be used. This is mostly because the physical point of the countersink can be from 0.005 to 0.020 inches short of the countersink's theoretical angle vertex. For example, the programmer will calculate Z axis travel by setting the length of travel from the countersink's theoretical vertex. Consequently, if the tool is preset to length with an indicator on the physical point, the vertex will be too deep. This will result in an oversized countersink.

As mentioned earlier, countersinking can be a programming problem on an N/C machine. It is important to remember that, when calculating and programming various countersinks, there is a difference between the theoretical vertex and the actual tool point.

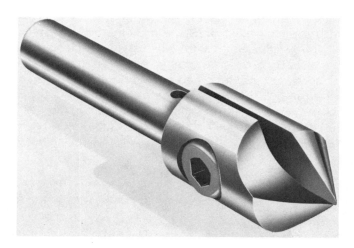

FIGURE 11-14
Standard, single-flute, nonpiloted countersink (Courtesy of Sharpaloy Division, Precision Industries, Inc. Centerdale, R.I.

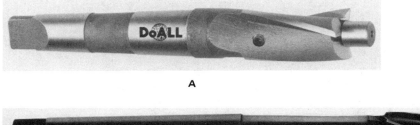

FIGURE 11-15
Typical piloted counterbores (A, Courtesy of DoAll Company; B, Courtesy of Cincinnati Milacron, Inc.)

Counterboring operations typically are done with a three- to eight-fluted counterbore, figure 11-15. These tools generally have about a ten-degree helix and are available with either tapered shanks or small diameter shanks ideally suited for straight collet toolholders. Counterbores are designed with fixed or removable pilots to produce counterbores concentric with the drilled hole. With the repetitive positioning accuracy of most N/C machines, however, the need for piloted counterbores is often eliminated. Standard counterbores may be removed from the tooling for machining centers, except in long-reach applications.

Most shops with machining centers will probably have a stock of end mills, the diameters of which will range from 3/16 inch to 2 inches. With no need for pilots, it is good machining practice to make the necessary counterbores from end mills. Doing so will yield a large variety of counterbores with almost infinite size availability, and the faster helix means greater feed rates and faster chip removal.

Some advantages of using end mills as counterbores are: reduced tool inventory; lower tool cost per piece; and ease in producing a spotface or counterbore on a rough or angled surface.

HIGH-TECHNOLOGY TOOLS

No discussion of N/C tooling is complete without mentioning some of the new innovations in cutting tool technology. Recent advances have provided modern metal working with carbide insert drills of numerous types and styles, along with titanium-coated and ceramic inserts for both lathe and spindle tooling. Some combination drills, similar to that shown in figure 11-16, can be used for facing and turning operations as well as for drilling a hole from solid on an N/C lathe — all with the same tool! In addition, cutting speeds and feeds are greatly increased, thereby improving and maximizing productivity levels.

FIGURE 11-16
Carbide insert drill-face-turn tool (Courtesy of The Valeron Corporation)

FIGURE 11-17
High-technology car-
bide insert drill (Cour-
tesy of The Valeron
Corporation)

The high-technology indexable insert drilling bars, figure 11-17, were designed to replace conventional twist drills and spade drills. These drills are capable of running up to ten times faster, depending upon material type, because they run at coated carbide speeds and feeds. The indexable insert drilling bar uses multiple-edge, two-sided indexable inserts in most cases. Use of these indexable insert drills reduces cycle inhibit time because only the inserts need to be replaced, not the entire drill. The inserts can be indexed while still at the machine, and the use of inserts eliminates the need for tool resharpening.

Most carbide insert drills have other features which make them extremely advantageous over conventional twist drills. Indexable insert drills provide increased web thickness. This gives them the strength to handle high penetration rates. The larger shank diameters provide added rigidity and help avoid chatter. In drilling where a finish bore operation is required, a hole can be drilled very close to the final size desired; consequently, subsequent semifinish boring operations can be eliminated.

These types of drills, besides adding rigidity, operate at substantially higher metal removal rates than conventional high-speed twist drills or spade drills. For this reason, good machining practices are mandatory. Flying chips, for example, create a danger to the operator; a safety shield is usually required. Coolant is also necessary to cool the cutting edge and to backflush chips. Sufficient coolant must be applied continuously on these tools. This is due to the higher chip removal rates, speeds, and feeds. In addition, drilling with carbide insert drills develops high thrusts. If the setup is not rigid, the forces will create chatter and side loading. This can result in broken inserts and damage to the drilling bar.

High-technology tools are constantly being researched and tested in modern manufacturing applications. As the quest for better and more efficient methods of production increase, so will better and more efficient cutting tools.

FIXTURING

Proper fixturing is also extremely important to successful N/C machining. A poorly designed or manufactured fixture often causes problems at the machine, holds up production, or produces much waste. If proper design considerations are applied, these costly and time-consuming problems can be avoided.

The basic function of a typical fixture, figure 11-18, is to locate and secure the part for succeeding machining operations. This involves initial set-up time for the workpiece to be loaded and clamped in the fixture for machining. Loading and unloading constitute an important part of the nonproductive cycle time of each part. By simplifying this operation, more parts can be produced each hour. Therefore, fixtures should be designed to reduce part set-up time.

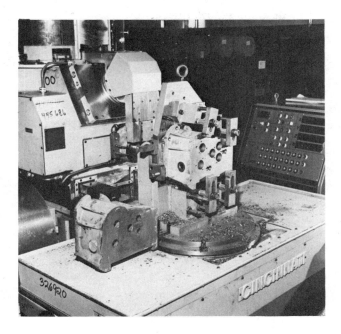

FIGURE 11-18
A typical N/C machining center fixture (Courtesy of Cincinnati Milacron Inc.)

N/C machines normally allow smaller batch lots, thereby reducing part inventories. Consequently, the fixture will be set up on the machine and used more frequently. This will warrant much consideration of design to reduce the set-up time and simplify the process of locating and securing the fixture to the machine table.

The accuracy of all parts requiring special fixtures depends on the workholding device regardless of how well the part is programmed or processed. Money spent taking the time to cover all aspects of locating and holding the workpiece accurately and securely will pay dividends when part and fixture reach the shop floor. This time and money should be spent in the initial design phase before production begins.

The advantages of sound, economical design and accurate manufacture and assembly of fixtures are:

- reduced fixture-to-machine and part-to-fixture set-up time.
- consistency of part accuracies.
- reduced errors and inaccuracies in part location.
- decreased cost per part.
- reduced inspection time.
- minimal fixture modifications or rework.
- faster and easier N/C tape prove-out.

Another important aspect of fixture design is that of clamping the workpiece. Clamps should always be placed as close to the support locations as possible. Placing unsupported clamps at any convenient location on the fixture could mean distortion of the part under clamping pressure. Clamp positions, for ease of loading and unloading, should also be given a priority. Clamps must avoid blocking hole locations and milling cuts that may interfere with part processing. Toolholders should be considered, particularly when the cutter is engaged in the workpiece, as long as they will not interfere with axis movements.

Clamping should become increasingly automated with larger part volumes. The types of clamps that should be used are pneumatic- and hydraulic-actuated clamps, toggle clamps, and cam-actuated clamps. Regardless of the type to be used or the number of parts to be processed, fixtures should always be designed with safety in mind. The operator should always be able to reach easily all clamps and any adjusting screws that may need attention. All sharp corners and projections should be minimized, and chip removal should be easy. The perfectly designed fixture is also the safest for those who must use it.

One overlooked aspect of fixturing that sometimes becomes evident after the fixture is in production is part orientation. Fixtures should always be designed to prevent incorrect loading and part orientation. Foolproofing of fixture design to prevent incorrect loading is critical; carelessness can occur when work is so repetitive. To foolproof the fixture design, there should be only one way the part can be located and clamped. This plan may take time to develop but it will alleviate scrapped parts and broken tools.

Several ideas must be considered regarding the positioning of the part and fixture in relation to the machine table.

- Will all the tools reach?
  The maximum and minimum tool lengths should be considered to determine optimum positioning of the fixture-to-table and part-to-fixture relationships.

- Is the fixture in a foolproof position to the machine table?
  Much time and effort is often spent designing a foolproof part-to-fixture relationship, and the fixture is loaded and clamped 90 or 180 degrees out of position. This can have expensive and hazardous consequences if the fixture is not marked to orient it to the machine table.

- Is the fixture designed and placed on the machine table to support efficient part processing?
  Sometimes inefficient N/C moves are needed to avoid poorly designed and placed locating pads and clamp assemblies. If the machine has a rotary index table, the part-to-fixture and fixture-to-table relationships should ideally be positioned over the center of rotation. This should provide an equal distance from the center of rotation to each face to be machined. It should also promote programming simplicity in X, Y, and Z axes.

SOME SOUND TOOLING PRACTICES

Regardless of the N/C machine to be used, types of cutting tools employed, or method of holding and locating the workpiece, sound cutting guidelines and qualified tooling practices must be followed if a machine is to be used to its greatest capacity.

There is no easy analysis to be made for any machining operation. Material characteristics and a wide assortment of other conditions make it difficult to formulate general solutions to tooling problems. Each situation requires individual analysis after careful consideration of all the variables. There are, however, some general N/C tooling practices, reminders, and guidelines which warrant discussion.

1) *Check all tools* before they are used. This includes the cutting edges and tool body as well as holders, extensions, adapters, etc. They should all be in perfect order and able to act as a total tool assembly.

2) *Select the right tool* for the job and use the tool correctly. Tools should be able to machine the workpiece to the desired accuracy. In purchasing cutting tools and toolholders, examine cost per piece part produced as well as the cost of the tooling package. In many cases, bargain tools cost more per part produced.

3) *Always choose the shortest drill length* that will permit drilling the hole to the desired depth. The smaller the drill size, the smaller the allowable error; as drill size increases, the allowable error progressively increases.

4) *Check and maintain correct cutting feeds and speeds for all tools.* This idea can spell success or failure for any N/C installation. Optimum feeds and speeds may not always be achievable, depending on machinability and material characteristics, rigidity of setup, etc.

5) *Understand machine and control capabilities.* Additional tools are often purchased because built-in machine/control capabilities such as contouring are not used effectively. Most controllers sold are of the contouring type, but qualified personnel must still be able to recognize where and how their capabilities can be used to alleviate tooling problems.

6) *Care for tools properly.* This includes perishable tools, holders, drivers, collets, extensions, etc. They represent a sizable investment and should be adequately stored and reconditioned when required.

7) *Watch and listen for abnormal cutting tool performances.* Attention paid to actual metal-removal processes can often prevent tool breakage problems, scrapped parts, and rework. Chatter and other vibration abnormalities can be corrected if detected early enough.

8) *Use correct tap drills.* Often the wrong drill is used for a certain size tap, and taps are sometimes broken as a result of negligence and incorrect tool selection.

9) *Holes to be tapped should be deep enough, free of chips, and lubricated prior to tapping.* Insufficient or incorrect lubrication can cause tap breakage, oversized threads, and poor surface finish.

10) *Use end mills as counterbores* where possible. They reduce inventory, produce a lower tool cost per piece, and produce a spotface or counterbore on a rough or angled surface more easily.

11) *The length-to-diameter ratio should not be exceeded on boring bars.* The boring bar length should never exceed four times the bar diameter. Failure to comply with this rule could result in chatter and inaccuracies in hole size.

12) *Select standard tools* whenever the operating conditions allow. Standard tools are less expensive, readily available, and interchangeable.

13) *Select the largest toolholder shank* the machine tool will allow. This will minimize deflection and reduce the tool overhang ratio.

14) *Select the strongest carbide insert* the workpiece will allow. This will increase overall productivity and lower the actual cost per insert cutting edge.

15) *Use negative rake insert geometry* whenever the workpiece or the machine tool will allow for it. This will double the cutting edges, provide greater strength to the insert, and dissipate the heat.

16) *Select the largest insert nose radius,* but the smallest insert size, that either the workpiece or machine tool will allow. The smallest size insert will be less expensive; using the largest insert nose radius will improve finish, dissipate heat, and provide greater strength.

17) *Select the largest depth of cut and the highest feed rate* that either the workpiece or machine tool will permit when using carbide inserts. This will improve overall productivity and have a negligible effect on tool life.

18) *Know the workpiece material and its hardness.* It is essential to have a thorough knowledge of the material's machining characteristics. If little is known, start at the lowest given cutting speed for that particular material and gradually increase the speed until optimum results are obtained.

19) *Select the cutting speed in relation to the physical properties of the workpiece.* The Brinell hardness number usually gives a good indication of the material's relative machinability.

20) *Increase cutter life by using lower cutting speeds and increasing the feed rate* to the limits allowed by the results desired, the setup rigidity, and the strength of the tool.

21) *For maximum cutter life, the feed should be as high as possible.* Doubling the feed (measured as chip per tooth) will double the stock removal per unit of time without appreciably decreasing cutter life.

22) *Use low cutting speeds for long cutter life.* However, soft, low-alloy materials can be machined at high cutting speeds without seriously affecting cutter life.

23) *Excessive cutting speeds generate excessive heat, resulting in shorter cutter life.* Chip and cutter tooth discoloration are good indications of excessive cutting speeds.

24) *Be aware that when the tool's cutting edges quickly become dull,* without chip or tooth discoloration, either the workpiece is very abrasive or there is a high resistance to chip separation. The cutting speed must be reduced.

25) *Use a sharp corner milling cutter only when* the job calls for milling to a sharp corner. Otherwise, use a cutter with a large corner chamfer. Greater stock removal with lower horsepower requirements will result.

26) *Climb milling will allow higher cutting speeds.* In addition, it will improve finish and lengthen cutter life.

27) *Coarse tooth end mills are preferred for roughing cuts.* Although some operators prefer fine tooth end mills for finishing cuts, it is possible to obtain finishes by increasing the cutting speed and decreasing the feed (chip load per tooth) of the coarse tooth end mill. Thus, the same end mill can be used for roughing and finishing.

28) *Direct cutting forces against the solid portion of both the machine and the fixture.* If work is held in a vise, direct the cutting forces against the solid jaw.

29) *Use coolants to get maximum cutter life* and to permit operating at higher cutting speeds. While coolant is not normally used when milling cast iron, by applying a jet of air as a coolant finish can be improved and cutter life can be lengthened.

30) *Fixture design should be simple* and standard components should be used whenever possible.

31) *The fixture and part should be located positively.* Rough, nonflat parts should be supported in three places and located on tooling holes if possible.

32) *Fixtures and parts should be readily accessible and movable* during any part of the machining operation and replaceable in exactly the same position.

33) *Fixture design should be simple* and *foolproof* for part loading to fixture and fixture loading to table.

34) *Always keep clamps close to fixture supports* in fixture design, and consider safety when designing for load and unload capabilities.

These tooling practices will avoid potential tooling problems on N/C machines. Consideration should be given to these points for successful use of N/C machines.

REVIEW QUESTIONS

1. Explain, in your own words, why tooling considerations are so important to success on an N/C machine.
2. What is the most important criterion when selecting the common twist drill for an N/C machine?
3. Why is center or spot drilling important to hole location?
4. When should spade drills be used instead of common twist drills? Why?
5. Discuss the advantages of multiple-diameter, multiple-land tools over single-diameter, single-land tools.
6. What are the most common causes of tap breakage on N/C machines? How can they be avoided?
7. What are hand taps? When should hand taps be used?

8. What is the primary difference between spiral pointed or gun taps and other taps? When should spiral pointed or "gun" taps be used?

9. When should spiral-fluted taps be used? What are their primary advantages?

10. Explain how fluteless taps work. What is the most important point to remember when using fluteless taps?

11. Discuss the importance of reaming on an N/C machine. Why are floating toolholders usually not required for reaming operations on N/C machines?

12. What is meant by the length-to-diameter ratio (L/D) in boring? What is the general rule to follow when applying this ratio for boring bars?

13. Describe how end mills and a machining center's contouring capability might be used in place of additional boring bars.

14. When should end mills be used for counterboring? Describe the main advantages of using end mills rather than ordinary counterbores.

15. What are some advantages of high-technology indexable insert drilling bars?

16. Discuss other factors which must be considered when using high-technology tools (e.g., flying chips, coolant usage, thrust, setup rigidity, etc.).

17. Why is fixturing so important to the success of an N/C machine?

18. What are the advantages of sound economical fixture design?

19. What is meant by foolproofing a part to fixture and fixture to machine table?

20. Why is it important to inspect thoroughly all cutting tools and fixtures before use?

CHAPTER 12

The Future of
Numerical Control

OBJECTIVES   After studying this chapter, the student will be able to:

- Identify additional computer applications and their role in technological advancement.

- Define the terms CAM, CAD, and CAPP, and discuss their importance to additional improvements in productivity.

- Explain the concept of computer graphics and its impact on engineering and manufacturing.

- Develop a conceptual understanding of manufacturing systems and their effect on future industrial capabilities.

BEYOND THE PROCESSOR LANGUAGES

In the future, what will take the place of our current processor languages such as APT? It appears that the successes which APT, its supporting post-processors, and other common N/C languages have thus far attained will stabilize or continue to increase gradually. Some changes from manual to computer techniques have occurred only recently and are not about to be quickly replaced. However, some improvements for the general processor languages are under way at present and in wide use in some companies. These changes will affect the overall method of part processing through computer application.

One of these changes concerns the multitude of postprocessors written to support APT and other languages. Most postprocessors are written for hard-wired controls, and do not contain some of the advanced routines such as stored parametric subroutines and repeatable patterns of modern CNC units. These sophisticated techniques affect the larger processor languages by eliminating from the actual program routines which may now be contained in the CNC unit. The recent developments in CNC technology are primarily due to minicomputer and microprocessor advancements. These new capabilities reduce the need, in some cases, for large, general processor applications. In addition, they further reduce computer application requirements by making manual part programming justifiable and efficient.

The real effect on APT and its derivatives has been the advent of CAD/CAM (Computer Aided Design/Computer Aided Manufacturing) systems. It was N/C that sparked the initial computer application of some complex workpieces. Computers are now being used to control all types of machines, processes, and systems. This integrated control of the design function, incorporated with manufacturing usage, is commonly referred to as CAD/CAM. The amount of computational and generative power contained in CAD/CAM systems is phenomenal and their use is ever increasing. However, APT and other processor languages have proven incompatible in many cases with some *CAD* systems' geometry. This is due to the manner in which detailed part geometry is stored within the computer. The computer, in turn, creates a new demand for N/C tool path information from a stored but usable engineering data base.

APT and its derivatives are relatively safe from obsolescense for the present, but work continues to discover better ways to use CAD/CAM and computer graphics for manufacturing purposes.

COMPUTER AIDED DESIGN

CAD/CAM is used extensively and implies a very diverse spectrum. CAD/CAM systems can range from minicomputer-based systems to large mainframe computers. CAD/CAM allows for detail parts and assemblies to be designed in an interactive environment with design geometry, being stored in a central data base for manufacturing engineering retrieval and part processing.

CAD has its base in computer graphics. These computer systems contain functional architecture for the design and drafting aspects of component parts and assemblies on a computer graphics CRT (cathode ray tube) display terminal. The principles of computer graphics are used to create lines, surfaces, solids, intersections, and curved surfaces. In simpler terms, computer graphics are systems that create, transform, and display pictorial, descriptive, and symbolic data. Graphics terminals, figure 12-1, are substitutes for conventional drawing boards whose part drawings can be stored in a central engineering computer data base. Conventional drawings and documents normally recorded on paper are cumbersome to work with, difficult to retrieve, and easily misplaced and misfiled.

Computer graphics, in contrast, have become a blessing for the CAD engineer. CAD capabilities range from using computers to create drawings to performing isolated calculations and compiling a bill of materials. Recorded images can range from a simple straight line to a multicolored pictorial representation of a three-dimensional assembly. Some images feature sculptured surfaces and moving parts, with shading and perspective to promote depth visualization. These descriptive geometric representations can then be rotated and viewed like an object in space, giving the designer total part viewing capabilities.

FIGURE 12-1
A typical computer graphics terminal (Courtesy of Cincinnati Milacron Inc.)

CAD systems are highly interactive and user-oriented. In many cases, they use construction techniques familiar to the conventionally trained draftsman. An interactive system provides immediate responses to the user's instructions, changes, or additions through the use of a light pen, panel buttons, etc. This greatly enhances a designer's creativity and conceptual thinking. Before CAD, the designer would sit in front of a drawing board; now that person sits in front of the CRT and creates images on the screen through the use of a light pen and panel buttons. The designer can also add and reproduce dimensions and symbols and manipulate the constructed images in a variety of ways never before possible with conventional paper and pencil. Once the engineer/designer has arrived at the final version or design, the image, because it is based on mathematical coordinates and entities stored within the computer, can be transmitted to peripheral devices for printing, plotting, etc. At this point, the *hard copy* is generated through computer instruction and passed on for manufacturing use.

One of the most important aspects of a CAD graphics system is that once the final part image is created, it can be stored in the engineering computer data base. This makes it readily accessible for viewing by other engineering personnel and ultimately for manufacturing use. From a design point of view, the storage of images greatly enhances and aids compatibility and interference visualization. Views can be merged, stacked, and rotated for assembly clarification without having to draw the assembly on paper and then find some interference factors necessitating redesign. Replication of details is also possible. A designer may construct details such as a fastener or a bracket only once and then replicate and locate it as necessary, making the geometry of a part available to other users. A library of standard symbols can be stored in the system and called up by users as needed. Most engineering data base systems provide both data management and data protection. The system controls the deletion of data and protects against unauthorized changes to drawings. Terminal and usage activity is monitored and recorded on a regular basis.

In addition to providing interactive design geometry capabilities, most computer graphics systems also provide advanced and powerful software programs which can analyze and test a design before any prototype parts are manufactured. Internal routines, such as finite element analysis, allow the engineer to calculate and predict patterns of stress and strength as well as other critical factors such as volume and weight.

All of the factors mentioned in this section are innovative, developing, and inspiring tools for the design engineer. However, some graphics systems are slanted toward the design function and less toward the manufacturing function. In order to complete the CAM function of a total CAD/CAM system, the CAD portion of the engineering data base must be accurate, accessible, and usable.

COMPUTER AIDED MANUFACTURING

In general, CAM refers directly to the manufacturing use of previously designed and stored CAD data. Most computer graphics systems contain a graphically oriented CAM part which allows a manufacturing engineer to retrieve selective part geometry. Tool paths are determined using the same light pen, control panels, and push buttons that were previously used in the part design phase (figure 12-1). The cutter path is routed around part contours or to the specific hole locations on the graphics terminal through cutter animation. The calculated tool paths are then verified and the computed coordinates are transmitted to an APT-type postprocessor for tape data generation. Thus, an N/C programmer can sit in front of a graphics display terminal, just like the designer, and watch the calculated and verified tool paths occur in relation to the actual part. As the cost of these systems reduces further and they become more widely used, their overall impact on conventional programming methods will become more significant.

For the most part, APT and its derivative processor languages are nongraphic, batch-oriented systems. The real advantage from a CAM point of view is that detail part geometry, if entered correctly, does not have to be recreated through the traditional part definition statements. The geometry already resides in the engineering data base. Thus, the question arises — why recreate what should already be accessible? Being able to capture stored and accurate part geometry on an engineering data base for manufacturing use is the most important aspect of a true CAM system. Improved part design accuracy further reduces the time for verifying N/C tapes. In addition, it reduces tool path attempts and scrapping. Part programming, using N/C graphics, enhances the part programmers' ability to follow visually the tool path by obtaining a three-dimensional view of cutter clearance planes, retract planes, depth planes, and clamps, fixtures, and casting clearances. Tape prove-out can be accomplished on the graphics display terminal rather than idling the N/C machine and operator on the shop floor.

The area under constant development in CAD/CAM systems is the incompatibilities of data base structures and the necessary link between CAD and CAM. Further, cost is a prohibitive factor when purchasing a CAD/CAM graphics-oriented system, even though the computational, analytical, and manufacturing features justify themselves.

Once the tool path data is calculated through cutter animation on the graphics terminal and the N/C tape data is generated, the N/C machine will be ready to accept information. In a later section, the use of N/C data from the CAM system in an automated systems approach such as DNC (direct numerical control) or FMS (flexible manufacturing system) will be discussed.

COMPUTER AIDED PROCESS PLANNING (CAPP)

Process planning involves creating detailed plans of the manufacturing steps and equipment necessary to produce a finished part. Workpiece requirements call for detailed analyses and accurate descriptions prior to the actual manufacturing process. A large assortment of machines and operations, as well as many different workers with a variety of skills, may be involved in the production of a specific part.

The computer lends itself well to the vital process planning function with two different approaches. One is called the "variant" or "similar part" method of process planning and the other is "generative." Both will produce an equal or accurate process plan, but most computer applications are of the variant type. This is because the software is easier to develop and new process plans are based on previous ones.

The variant method had its beginnings with the group technology concept, along with parts classification and coding systems. Group technology is a manufacturing philosophy based on the idea that similarities occur in the design and manufacture of component parts. These parts can be classified into groups, or families, if the basic configurations and attributes are identified. A reduction in expenses can be achieved through the structured classification and grouping of parts into families based upon engineering design and manufacturing similarities.

Using the variant method, CAPP groups families of parts by a structured classification and coding plan. All previously processed parts are coded using this method. The parts are then divided into part families, such as rotational and bar and rail, based on general configuration. A standard order of operations or sequences is stored on the computer for each part family. When a new part is ready for planning, the classification or group technology code for the new part is used to compare and retrieve the standard process plan for that part family. Editing capabilities further enable the process planner to alter the standard order of operations for final refinement. The completed process plan is then stored on the computer data base by part number.

In generative process planning, the parts are again broken into part families, and a detailed analysis is made for each part family to determine individual part operations. This type of system develops the actual operation sequence based on the part geometry, usage requirements, material size and configuration, and available equipment. The generative approach creates process planning logic for the part family groupings. The logic is then stored internally as a decision model. As new workpieces require process planning, analyses must be conducted to determine and compare the features incorporated in the decision model with those on the actual part. The family part decision model is then retrieved, and a routing sheet is generated by processing the decision model with the new workpiece attributes.

A generative system must be driven by much more elaborate and powerful software than a variant system. Development and optimization work continues on the variant and generative approaches to process planning. Both systems, however, build the needed decision-making logic and planning ability into the computer rather than rely on a decreasing experience level in the process planning work force.

MANUFACTURING SYSTEMS

The concept of a manufacturing system can range from a DNC system to FMS. Numerical control, computer numerical control, and direct numerical control are part of the total manufacturing system. They contribute to the automated manufacturing systems of today and tomorrow.

DNC installations have a computer interface to stand-alone machines. The DNC installation offers several operating advantages as previously discussed. We already studied that the tape reader is bypassed for downloading N/C information from the computer to the MCU. We also know that a program in computer storage is easily accessible for programmer interaction of revision and editing. However, the same computer that directs the operation of a machine tool can also be used for auxiliary purposes such as machine downtime recording, performance tabulation, real time machine status, and other items of interest to management. Gathered information may then be kept in the controlling computer's memory for retrieval and study. In addition, advanced DNC units can be used to sense the operating conditions of various machines through CNC feedback to the host computer and output error diagnostics or initiate corrective action. The computer, through the evolution of N/C and DNC, makes it possible to create versatile systems, capable of producing different parts and automatically adapting to different mixes and variations of part types and lot sizes.

A manufacturing system can be described as automation that combines processing machinery, tooling, and material handling equipment. The success of manufacturing systems depends on astute integration of concept, design, software, and quality.

CNC machines fit well into a manufacturing system's concept as they are primarily used in the mid-volume production range and are capable of many operating tasks in a DNC environment. In order to use CNC machines to their fullest capacity, the spindles must keep turning to increase chip-making time. This can be accomplished through part program storage, automated handling devices, and sophisticated monitoring systems. A system of this type can provide random manufacturing of different workpieces without individual machine setup time, thereby maximizing spindle time, decreasing in-process inventory, and reducing lead time. Figure 12-2 shows a typical manufacturing system.

A manufacturing system can be described using six generic headings:
1) description and design of parts to be produced;
2) configuration and layout;
3) standardization of tools and preset;

FIGURE 12-2
A typical manufacturing system (Courtesy of Cincinnati Milacron Inc.)

4) prior setup (load and unload);
5) transport capabilities; and
6) operation and control.

DESCRIPTION AND DESIGN OF PARTS TO BE PRODUCED

The ideal beginning of an efficient manufacturing system is engineering. The type and families of parts to be processed will determine, in many cases, the number and types of machines in the system, tooling and fixturing requirements, load, unload, and transport capabilities as well as the actual system design and layout. Parts may need to be redesigned to reduce the large proliferation of tool inventory and to standardize design considerations so future products will fit the system.

CONFIGURATION AND LAYOUT

Annual part usage requirements and average lot sizes play a large role in determining system structure and layout. Transport devices must have easy and accessible entry and exit to loading or queuing stations and CNC machines. Part scheduling, potential interference problems, cutting tool replacements, and maintenance considerations must be thoroughly planned and approved before system implementation begins.

STANDARDIZATION OF TOOLS AND PRESET

Tools and preset of tool assemblies can make or break a manufacturing system. It is important to maintain a standardized set of resident cutting tools with several backups or replacements as tools become worn, chipped, or broken. CNC units can monitor tool usage in addition to offsetting each tool to an optical sensor to check for tool breakage. Action can then be initiated to replace broken or worn tools with an identical tool preassembled and waiting at the machine. This minimizes machine spindle idle time.

PRIOR SETUP (LOAD AND UNLOAD)

One of the most costly problems of stand-alone N/C machines is that the spindle is idle while the operator replaces the part or sets up for a new workpiece. In a system's concept, parts should be set up on individual pallets in a part staging area prior to machining and left in queue until called for by the host computer. Finished parts are transported to the same off-line area for unloading while a new pallet with an unmachined part is delivered to the machine for processing. Off-line loading and unloading greatly increases spindle time and decreases machine idle time.

TRANSPORT CAPABILITIES

A true manufacturing system must provide and maintain an unmanned, automated process for material movement to and from the various machines. Some material carriers traverse back and forth on a track, accepting and delivering parts on pallets via computer command. Others follow energized wires embedded in the surface of the floor. Transport carriers provide horizontal transportation of discrete loads of material between specific locations. They also provide an essential element to automated manufacturing systems that cannot be taken for granted. Automated movement to and from the various work stations in a timely fashion represents a sizable portion of productivity improvement potential.

OPERATION AND CONTROL

The integrated and sophisticated host computer software is the heart of a manufacturing system. In addition to directing the DNC aspect of the machines contained within the cell and storing all of the part programs, the host computer must detect part identification, monitor and detect tool wear or breakage, handle inspection information, initiate pallet and material movements, and respond to signals indicating malfunctions, among other duties. The system computer also takes care of work scheduling and collection of machining results. By entering data such as work lot size, machining time, and machining priority, schedules for each machine and work flow can be prepared and reliably predicted. Machining centers generally form the core of a system. Whether the system achieves the desired results depends on individual machine uptime and how efficiently the machining centers are run.

In a regular process-oriented, stand-alone machine shop, it may take weeks to process a complete part. Approximately 95% of this time is waiting, part inactive time. By reducing the waiting times, the use of manufacturing systems will shorten the cycle time of parts produced and subsequently minimize in-process inventory. In-process inventory is one of the most costly items in metal-working operations.

Manufacturers today must find ways to overcome internal inefficiencies and free up working capital by increasing machine use and through-put, and decreasing the in-process inventory. The market for both dedicated and flexible manufacturing systems will accelerate and continue to grow in the future. They will be seen as the best way to increase productivity and return on investment.

Obviously, more detail is required in a manufacturing system than the conceptual overview presented here. It is important to realize that numerical control will continue to play an important role in automated manufacturing systems.

LOOKING AHEAD

What began as an idea created in the early 1950s by John Parsons has developed into a manufacturing concept that has revolutionized the metal-cutting industry. Numerical control has advanced from a stand-alone method of its own to a subset of a larger and much broader CAD/CAM industry.

Numerical control is bound to hold major improvements in the area of part programming. Large installations have tremendous engineering CAD-developed data bases which manufacturing must use for CAM activity. Development will continue and prices will become more affordable due to advanced electronics and sophisticated software. But what about the smaller job shops? Additional work and development will continue on N/C languages used for time sharing and basic processing in the immediate future. These languages will consist of commands that are derived from human language. Consequently, the programming process will become easier to understand and will have a shorter learning curve as new programmers enter the field. Thus, programming skill will become less coding-oriented and more process-oriented.

Additional improvements will take place in the area of machine tool controls. N/C units will possess high-level user graphics systems, thereby permitting communication between the operator and the MCU for cutter path verification and correction. The N/C tape will eventually be eliminated. Data transmission devices, such as bubble-memory cartridges, and other sophisticated media yet to be developed will take its place. Data can be transmitted directly to the machine control, stored, used, and updated without constantly generating and punching new tapes.

Unmanned machining operations, using robot applications, similar to that shown in figure 12-3 will also play a significant role in the future of manufacturing and numerical control. This is due to the following:

- A declining percentage of United States work force choosing careers in manufacturing.
- Increased production pressure from foreign competition.
- Internal inefficiencies that create costly in-process inventory and tie up working capital.
- Increased availability of robot systems for part loading and unloading.
- Significant advancements in electronics and microprocessor technology.
- The need for predictable and dependable productivity levels.

Although other factors may also affect the quest toward unmanned machining operations, the future clearly indicates that the United States will embark on an ambitious program, strongly committed to N/C technology, to rehabilitate manufacturing facilities and thus improve industrial productivity.

The N/C part programmer will always be a vital element in the manufacturing process. The demand for experienced part processing and part

FIGURE 12-3
An unmanned machining operation (Courtesy of Cincinnati Milacron Inc.)

programming personnel will grow. Emphasis will be placed on total job processing rather than actual part programming.

REVIEW QUESTIONS

1. Define CAD/CAM. How does CAD/CAM relate to computer graphics?
2. Why is the CAD engineering data base so important to CAM users?
3. What are the primary advantages of a CAD/CAM system?
4. How does a CAM system assist in eliminating machine spindle idle time, tape prove-out, and scrapping?
5. What does CAPP mean? What is its overall impact on manufacturing?
6. Explain the difference between variant and generative process planning.
7. Briefly explain the six basic elements of a manufacturing system. Discuss the importance of each element.
8. How are computer graphics significant to future productivity improvement from CAD and CAM viewpoints?

APPENDIX A

EIA and AIA National Codes

PREPARATORY FUNCTIONS

G Word	Explanation
G00	Used for denoting a rapid traverse rate with point-to-point positioning.
G01	Used to describe linear interpolation blocks and reserved for contouring.
G02 G03	Used with circular interpolation.
G04	A calculated time delay during which there is no machine motion (dwell).
G05 G07	Unassigned by the EIA. May be used at the discretion of the machine tool or system builder. Could also be standardized at a future date.
G06	Parabolic interpolation.
G08	Acceleration code. Causes the machine, assuming capability, to accelerate at a smooth exponential rate.
G09	Deceleration code. Causes the machine, assuming capability, to decelerate at a smooth exponential rate.
G10 G11 G12	Normally unassigned for CNC systems. Used with some hard-wired systems to express blocks of abnormal dimensions.
G13 G14 G15 G16	Used to direct the control system to operate on a particular set of axes.
G17 G18 G19	Used to identify, or select, a coordinate plane for such functions as circular interpolation or cutter compensation.
G20 through G32	Unassigned according to EIA standards; however, they may be assigned by the control system or machine tool builder.
G33 G34 G35	Modes selected for machines equipped with thread-cutting capabilities and generally referring to lathes. G33 is used when a constant lead is sought. G34 is used when a constantly increasing lead is required, and G35 is used to designate a constantly decreasing lead.

G36 through G39	Unassigned.
G40	A command which will terminate any cutter compensation.
G41	A code associated with cutter compensation in which the cutter is on the left side of the work surface, looking in the direction of the cutter motion.
G42	A code associated with cutter compensation in which the cutter is on the right side of the work surface.
G43 G44	Used with cutter offset to adjust for the difference between the actual and programmed cutter radii or diameters. G43 refers to an inside corner, and G44 refers to an outside corner.
G45 through G49	Unassigned.
G50 through G59	Reserved for adaptive control.
G60 through G69	Unassigned.
G70	Inch programming.
G71	Metric programming.
G72	Three-dimensional circular interpolation-CW.
G73	Three-dimensional circular interpolation-CCW.
G74	Cancel multiquadrant circular interpolation.
G75	Multiquadrant circular interpolation.
G76 through G79	Unassigned.
G80	Cancel cycle.
G81	Drill, or spotdrill, cycle.
G82	Drill with a dwell.
G83	Intermittent, or deep-hole, drilling.
G84	Tapping cycle.
G85 through G89	Boring cycles.
G90	Absolute input. Input data is to be in absolute dimensional form.
G91	Incremental input. Input data is to be in incremental form.
G92	Preload registers to desired values. An example would be to pre-load axis position registers.
G93	Inverse time feed rate.
G94	Inches (millimetres) per minute feed rate.
G95	Inches (millimetres) per revolution feed rate.
G97	Spindle speed in revolutions per minute.

$\left.\begin{array}{l}\text{G98} \\ \text{G99}\end{array}\right\}$ Unassigned.

MISCELLANEOUS FUNCTIONS

M Word	Explanation
M00	Program stop. Operator must cycle start in order to continue with the remainder of the program.
M01	Optional stop. Acted upon only when the operator has previously signaled for this command by pushing a button. When the control system senses the M01 code, machine will automatically stop.
M02	End of program. Stops the machine after completion of all commands in the block. May include rewinding of tape.
M30	End of tape command. Will rewind the tape and automatically transfer to a second tape reader if incorporated in the control system.
M03	Start spindle rotation in a clockwise direction.
M04	Start spindle rotation in a counterclockwise direction.
M05	Spindle stop.
M06	Command to execute the change of a tool (or tools) manually or automatically.
M07	Turn coolant on (flood).
M08	Turn coolant on (mist).
M09	Coolant off.

$\left.\begin{array}{l}\text{M10} \\ \text{M11}\end{array}\right\}$ Automatic clamping of the machine slides, workpiece, fixture, spindle, etc. M11 is an unclamping code.

M12   An inhibiting code to synchronize multiple sets of axes, such as a four-axis lathe having two independently operated heads or slides.

M13   Combines simultaneous clockwise spindle motion and coolant on.

M14   Combines simultaneous counterclockwise spindle motion and coolant on.

$\left.\begin{array}{l}\text{M15} \\ \text{M16}\end{array}\right\}$ Rapid traverse or feed motion in either the +(M15) or –(M16) direction.

$\left.\begin{array}{l}\text{M17} \\ \text{M18}\end{array}\right\}$ Unassigned.

M19   Oriented spindle stop. Spindle stop at a predetermined angular position.

$\left.\begin{array}{l}\text{M20} \\ \text{through} \\ \text{M29}\end{array}\right\}$ Unassigned.

M31   A command known as interlock bypass for temporarily circumventing a normally provided interlock.

$\left.\begin{array}{l}\text{M32} \\ \text{through} \\ \text{M39}\end{array}\right\}$ Unassigned.

$\left.\begin{array}{l}\text{M40} \\ \text{through} \\ \text{M46}\end{array}\right\}$ Used to signal gear changes if required at the machine; otherwise, unassigned.

M47   Continues program execution from the start of the program unless inhibited by an interlock signal.

M48   Cancel M49.

M49   A function that deactivates a manual spindle or feed override and returns to the programmed value.

M50
through   Unassigned.
M57

M58   Cancel M59.

M59   A function which holds the RPM constant at its value when M59 is initiated.

M60
through   Unassigned.
M99

OTHER ADDRESS CHARACTERS

Address Character	Explanation
A	Angular dimension about the X axis.
B	Angular dimension about the Y axis.
C	Angular dimension about the Z axis.
D	Can be used for an angular dimension around a special axis, for a third feed function or for tool offset.
E	Used for angular dimension around a special axis or for a second feed function.
H	Unassigned.
I J K	Used with circular interpolation.
L	Not used.
O	Used in place of the customary sequence number word address N.
P	A third rapid traverse code or tertiary motion dimension parallel to the X axis.
Q	Second rapid traverse code or tertiary motion dimension parallel to the Y axis.
R	First rapid traverse code or tertiary motion dimension parallel to the Z axis or the radius for constant surface speed calculation.
U	Secondary motion dimension parallel to the X axis.
V	Secondary motion dimension parallel to the Y axis.
W	Secondary motion dimension parallel to the Z axis.

APPENDIX B

General Safety Rules
for N/C Machines

1) Wear safety glasses at all times.
2) Wear safety shoes.
3) Do not wear neckties, long sleeves, wristwatches, rings, gloves, etc. when operating machine.
4) Keep long hair covered when operating machine.
5) Make sure the area around the machine is well-lighted, dry, and as free from obstructions as possible. Keep the area in good order.
6) Never perform grinding operations near an N/C machine. Abrasive dust will cause undue wear, inaccuracies and possible failure of affected parts.
7) Do not use compressed air to blow chips from the part, machine surfaces, cabinets, controls, or floor around the machine.
8) When handling or lifting parts or tooling, follow company policy on correct procedures.
9) Work platforms around machines should be sturdy and must have antislip surfaces.
10) Wrenches, tools, and other parts should be kept off the machine and all its moving units. Do not use machine elements as a workbench.
11) Keep hands out of path of moving units during machining operations.
12) Never place hands near a revolving spindle.
13) Perform all setup work with spindle stopped.
14) Load and unload workpieces with spindle stopped.
15) Clamp all work and fixtures securely before starting machine.
16) When handling tools or changing tools by hand, use a glove or cloth. Avoid contact with cutting edges. Do not operate machine with gloves.
17) Use caution when changing tools, and avoid interference with fixture or workpiece.
18) Use only properly sharpened tools.
19) Clean the setup daily.
20) Avoid bumping any N/C machine or controls.
21) Never operate an N/C machine without consulting the specific operator's manual for that particular machine and control type.
22) Never attempt to program an N/C machine without consulting the specific programmer's manual for that particular machine and control type.
23) Electrical compartment doors should be opened *only* for electrical and/or maintenance work. They should be opened only by experienced electricians and/or qualified service personnel.

24) Safety guards, covers, and other devices have been provided for protection. Do not operate machine with these devices disconnected, removed, or out of place. Operate machine only when they are in proper operating condition and position.

25) Tools are made for right-hand or left-hand operation. Be sure spindle direction is correct.

26) Do not remove chips from workpiece area with fingers or while spindle is running. Use a brush to remove chips *after* the spindle has stopped. Clear chips often.

APPENDIX C

Useful Formulas and Tables

$$RPM = \frac{3.82 \times Cutspeed}{Diameter}$$

Feed (IPM) = (RPM) × (IPR)
(inches/minutes)   (RPM)(inches/revolution)

$$Feed \ (IPR) = \frac{feed(IPM)}{RPM}$$

- To find the circumference of a circle, multiply the diameter by 3.1416.

- To find the diameter of a circle, multiply the circumference by .31831.

- To find the area of a circle, multiply the square of the diameter by .7854.

- To obtain the circumference, multiply the radius of a circle by 6.283185.

- To obtain the area of a circle, multiply the square of the circumference of a circle by .07958.

- To find the area of a circle, multiply one-half the circumference of a circle by one-half its diameter.

- To obtain the radius of a circle, multiply the circumference of a circle by .159155.

- To find the radius of a circle, multiply the square root of the area of a circle by .56419.

- To find the diameter of a circle, multiply the square root of the area of a circle by 1.12838.

- To find the area of the surface of a ball (sphere), multiply the square of the diameter by 3.1416.

- To find the volume of a ball (sphere), multiply the cube of the diameter by .5236.

Trigonometry

$$c^2 = a^2 + b^2$$

$$c = \sqrt{a^2 + b^2}$$

$$a = \sqrt{c^2 - b^2}$$

$$b = \sqrt{c^2 - a^2}$$

$$\text{Sine} = \frac{\text{side opposite}}{\text{hypotenuse}} \qquad \text{Cosecant} = \frac{\text{hypotenuse}}{\text{side opposite}}$$

$$\text{Cosine} = \frac{\text{side adjacent}}{\text{hypotenuse}} \qquad \text{Secant} = \frac{\text{hypotenuse}}{\text{side adjacent}}$$

$$\text{Tangent} = \frac{\text{side opposite}}{\text{side adjacent}} \qquad \text{Cotangent} = \frac{\text{side adjacent}}{\text{side opposite}}$$

CUTTING SPEEDS
(Feet Per Minute)

MATERIAL	DRILL HSS	DRILL CARBIDE	REAM HSS	REAM CARBIDE	TAP HSS	COBORE HSS	COBORE CARBIDE	BORE HSS	BORE CARBIDE	MILLING HIGH SPEED Rgh.	MILLING HIGH SPEED Fin.	MILLING CARBIDE Rgh.	MILLING CARBIDE Fin.
Aluminum	200	350	175	300	90	180	300	300	600	240	300	500	1000
Brass – Soft	145	350	120	250	100	150	300	150	450	150	200	400	600
Brass – Hard	125	225	100	200	75	110	200	120	350	135	180	350	500
Bronze – Common	140	250	125	200	90	130	200	150	400	145	190	360	550
Bronze – High Tensile	60	200	50	175	40	55	180	85	300	70	90	200	280
Cast Iron – Soft 170 BHN	90	180	60	200	40	85	160	80	280	90	110	250	350
Cast Iron – Medium 220 BHN	60	140	45	125	30	55	130	55	255	70	90	200	300
Cast Iron – Hard 300 BHN	40	120	30	60	20	35	100	45	215	50	60	175	250
Cast Iron – Malleable	85	140	45	100	40	75	180	90	250	100	120	260	370
Cast Steel	60	120	50	100	40	60	180	70	200	50	80	225	380
Copper	75	250	50	125	40	70	200	95	350	90	150	220	400
Magnesium	250	500	180	450	150	200	450	400	1000	300	400	600	1000
Monel	50	100	35	90	20	45	90	50	110	60	80	180	240
Steel – Mild .2 to .3 Carbon	95	–	50	250	40	85	170	80	280	90	130	300	450
Steel – Medium .4 to .5 Carbon	75	–	45	200	35	60	120	80	220	70	85	210	400
Steel – Tool up to 1.2 Carbon	40	80	30	70	20	40	80	45	190	50	80	175	350
Steel – Forging	45	90	35	80	25	40	80	50	200	60	80	200	300
Steel – Alloy 300 BHN	60	120	40	115	35	60	120	70	250	60	80	250	350
Steel – Alloy 400 BHN	45	90	30	65	25	40	80	40	165	30	40	160	250
Steel – High Tensile to 40 R_c	35	70	30	60	20	30	60	40	150	40	50	120	150
Steel – High Tensile to 45 R_c	30	60	20	50	15	20	40	30	100	35	45	110	140
Steel – Stainless – Free Machining	55	110	35	100	25	50	100	50	150	40	60	200	400
Steel – Stainless – Work Hardening	30	60	20	50	15	30	60	40	90	30	50	180	300
Titanium – Commercially Pure	55	110	45	100	30	50	100	60	120	60	75	200	280
Zinc Die Casting	150	300	125	225	80	150	250	180	350	200	300	250	450

CONVERSION CHART — CUTTING SPEEDS TO RPM

CUTTING SPEEDS (left column) × TOOL DIAMETERS (in inches)

Small Tool Diameters

Cutting Speed	1/8	3/16	1/4	5/16	3/8	7/16	1/2	9/16	5/8	11/16	3/4	13/16	7/8	15/16	1	1-1/8	1-1/4	1-3/8	1-1/2	1-5/8	1-3/4
10	306	204	153	122	102	87	76	68	61	56	51	47	44	41	38	34	31	28	25	24	22
20	611	407	306	244	204	175	153	136	122	111	102	94	87	82	76	68	61	56	51	47	44
30	917	611	458	368	306	262	229	204	183	167	153	141	131	122	115	102	92	83	76	71	65
40	1222	815	611	489	408	349	306	272	245	222	204	188	175	163	153	136	122	111	102	94	87
50	1528	1020	764	611	509	437	382	340	306	278	255	235	218	204	191	170	153	139	127	118	109
60	1834	1222	917	733	611	524	458	407	367	333	306	282	262	245	229	204	183	167	153	141	131
70	2140	1426	1070	856	713	611	535	475	428	389	357	329	306	285	267	238	214	194	178	165	153
80	2445	1630	1222	978	815	700	611	543	489	444	408	376	350	326	306	272	244	222	204	188	175
90	2750	1833	1375	1100	917	786	688	611	550	500	458	423	393	367	344	306	275	250	229	212	196
100		2037	1528	1222	1020	873	764	679	611	556	509	470	437	408	382	340	306	278	255	235	218
120		2445	1834	1467	1222	1048	917	815	733	667	611	564	524	489	458	407	367	333	306	282	262
140		2852	2140	1711	1426	1222	1070	950	856	778	713	658	611	571	535	475	428	390	356	329	306
160			2445	1956	1630	1397	1222	1086	978	889	815	752	698	652	611	543	489	444	407	376	350
180			2750	2200	1834	1572	1375	1222	1100	1000	917	846	786	734	688	611	550	500	458	423	393
200				2445	2037	1747	1528	1358	1222	1111	1020	940	873	815	764	680	611	556	510	470	437
220				2690	2240	1920	1681	1494	1345	1222	1121	1034	960	897	840	747	672	611	560	517	480
240				2934	2445	2096	1834	1630	1467	1333	1222	1128	1048	978	917	815	733	667	611	564	524
250					2547	2183	1910	1697	1528	1389	1274	1175	1091	1020	955	850	764	694	637	588	546
260					2650	2270	1986	1765	1590	1444	1325	1222	1135	1060	993	883	795	722	662	611	568
280					2852	2445	2139	1900	1712	1556	1426	1316	1222	1140	1070	950	856	778	713	658	611
300						2620	2292	2037	1834	1667	1528	1410	1310	1222	1146	1020	917	833	764	705	655
320						2795	2445	2172	1956	1778	1630	1504	1397	1304	1222	1086	978	889	815	752	698
340						2970	2597	2308	2078	1889	1732	1600	1484	1385	1300	1155	1040	944	866	800	742
350							2674	2375	2140	1944	1783	1646	1528	1426	1337	1190	1070	972	891	823	764
360							2750	2444	2200	2000	1834	1693	1570	1467	1375	1222	1100	1000	917	846	786
380							2903	2580	2323	2111	1936	1787	1660	1550	1450	1290	1160	1055	968	893	830
400								2715	2445	2222	2038	1881	1746	1630	1530	1360	1222	1111	1020	940	873
450									2750	2500	2292	2116	1964	1834	1720	1530	1375	1250	1146	1060	982
500										2778	2550	2350	2180	2037	1910	1700	1530	1390	1273	1175	1090
550											2800	2586	2400	2240	2100	1870	1680	1530	1400	1293	1200
600												2821	2620	2445	2290	2040	1834	1667	1528	1410	1310

Large Tool Diameters

Cutting Speed	1-1/8	1-1/4	1-3/8	1-1/2	1-5/8	1-3/4	2	2-1/4	2-1/2	2-3/4	3	3-1/2	4	4-1/2	5	5-1/2	6	6-1/2	7	7-1/2	8
10	34	31	28	25	24	22	19	17	15												
20	68	61	56	51	47	44	38	34	31	28	25	22	19	17	15						
30	102	92	83	76	71	65	57	51	46	42	38	33	29	25	23	21	19	18	16	15	
40	136	122	111	102	94	87	76	68	61	56	51	44	38	34	31	28	25	24	22	20	19
50	170	153	139	127	118	109	96	85	76	69	64	55	48	42	38	35	32	29	27	25	24
60	204	183	167	153	141	131	115	102	92	83	76	65	57	51	46	42	38	35	33	31	29
70	238	214	194	178	165	153	134	119	107	97	89	76	67	59	53	49	45	41	38	36	33
80	272	244	222	204	188	175	153	136	122	111	102	87	76	68	61	56	51	47	44	41	38
90	306	275	250	229	212	196	172	153	138	125	115	98	86	76	69	63	57	53	49	46	43
100	340	306	278	255	235	218	191	170	153	139	127	109	96	85	76	70	64	59	55	51	48
120	407	367	333	306	282	262	229	204	183	167	153	131	115	102	92	83	76	71	65	61	57
140	475	428	390	356	329	306	267	238	214	194	178	153	134	119	107	97	89	82	76	71	67
160	543	489	444	407	376	350	306	272	244	222	204	175	153	136	122	111	102	94	87	81	76
180	611	550	500	458	423	393	344	306	275	250	229	196	172	153	138	125	115	106	98	92	86
200	680	611	556	510	470	437	382	340	306	278	255	218	191	170	153	139	127	117	109	102	96
220	747	672	611	560	517	480	420	374	336	306	280	240	210	187	168	153	140	129	120	112	105
240	815	733	667	611	564	524	458	407	367	333	306	262	229	204	183	167	153	141	131	122	115
250	850	764	694	637	588	546	477	425	382	347	318	273	239	212	191	174	159	147	136	127	119
260	883	795	722	662	611	568	497	441	397	361	331	284	248	221	199	181	166	153	142	132	124
280	950	856	778	713	658	611	535	475	428	389	357	306	267	238	214	194	178	165	153	143	134
300	1020	917	833	764	705	655	573	510	458	417	382	327	286	255	229	208	191	176	164	153	143
320	1086	978	889	815	752	698	611	543	489	444	407	349	306	272	244	222	204	188	175	163	153
340	1155	1040	944	866	800	742	650	577	520	472	433	371	325	289	260	236	216	200	186	173	162
350	1190	1070	972	891	823	764	668	594	535	486	446	382	334	297	267	243	223	206	191	178	167
360	1222	1100	1000	917	846	786	688	611	550	500	458	393	344	306	275	250	229	212	196	183	172
380	1290	1160	1055	968	893	830	725	645	580	528	484	415	363	323	290	264	242	223	207	194	181
400	1360	1222	1111	1020	940	873	764	680	611	556	510	437	382	340	306	278	255	235	218	204	191
450	1530	1375	1250	1146	1060	982	860	764	688	625	573	491	430	382	344	313	286	264	246	229	215
500	1700	1530	1390	1273	1175	1090	955	850	764	694	637	546	478	425	382	347	318	294	273	255	239
550	1870	1680	1530	1400	1293	1200	1050	934	840	764	700	600	525	467	420	382	350	323	300	280	263
600	2040	1834	1667	1528	1410	1310	1145	1020	917	833	764	655	573	510	458	417	382	353	327	306	287

TAPPING FEED RATES

THREADS PER INCH-TO-LEAD

TPI	LEAD	TPI	LEAD	TPI	LEAD
3	.3333	11-1/2	.0870	32	.0313
3-1/2	.2587	12	.0833	36	.0278
4	.2500	13	.0769	40	.0250
5	.2000	14	.0714	44	.0227
6	.1667	16	.0625	48	.0208
7	.1430	18	.0556	56	.0179
8	.1250	20	.0500	64	.0156
9	.1111	24	.0417	72	.0139
10	.1000	27	.0370	80	.0125
11	.0909	28	.0357		

Program feed rate = Lead of tap (inches) $\times$ RPM

CONVERSION CHART
(Based on 25.4 mm = 1")
Inches into Millimeters

Inches		M/M	Inches		M/M	Inches	M/M	Inches	M/M	Inches	M/M
1/64	.0156	0.3969	49/64	.7656	19.4469	34	863.600	82	2082.80	130	3302.00
1/32	.0313	0.7937	25/32	.7813	19.8437	35	889.000	83	2108.20	131	3327.40
3/64	.0469	1.1906	51/64	.7969	20.2406	36	914.400	84	2133.60	132	3352.80
1/16	.0625	1.5875	13/16	.8125	20.6375	37	939.800	85	2159.00	133	3378.20
5/64	.0781	1.9844	53/64	.8281	21.0344	38	965.200	86	2184.40	134	3403.60
3/32	.0938	2.3812	27/32	.8438	21.4312	39	990.600	87	2209.80	135	3429.00
7/64	.1094	2.7781	55/64	.8594	21.8281	40	1016.00	88	2235.20	136	3454.40
1/8	.1250	3.1750	7/8	.8750	22.2250	41	1041.40	89	2260.60	137	3479.80
9/64	.1406	3.5719	57/64	.8906	22.6219	42	1066.80	90	2286.00	138	3505.20
5/32	.1563	3.9687	29/32	.9063	23.0187	43	1092.20	91	2311.40	139	3530.60
11/64	.1719	4.3656	59/64	.9219	23.4156	44	1117.60	92	2336.80	140	3556.00
3/16	.1875	4.7625	15/16	.9375	23.8125	45	1143.00	93	2362.20	141	3581.40
13/64	.2031	5.1594	61/64	.9531	24.2094	46	1168.40	94	2387.60	142	3606.80
7/32	.2188	5.5562	31/32	.9688	24.6062	47	1193.80	95	2413.00	143	3632.20
15/64	.2344	5.9531	63/64	.9844	25.0031	48	1219.20	96	2438.40	144	3657.60
1/4	.2500	6.3500	1		25.4000	49	1244.60	97	2463.80	145	3683.00
17/64	.2656	6.7469	2		50.800	50	1270.00	98	2489.20	146	3708.40
9/32	.2813	7.1437	3		76.200	51	1295.40	99	2514.60	147	3733.80
19/64	.2969	7.5406	4		101.600	52	1320.80	100	2540.00	148	3759.20
5/16	.3125	7.9375	5		127.000	53	1346.20	101	2565.40	149	3784.60
21/64	.3281	8.3344	6		152.400	54	1371.60	102	2590.80	150	3810.00
11/32	.3438	8.7312	7		177.800	55	1397.00	103	2616.20	151	3835.40
23/64	.3594	9.1281	8		203.200	56	1422.00	104	2641.60	152	3860.80
3/8	.3750	9.5250	9		228.600	57	1447.80	105	2667.00	153	3886.20
25/64	.3906	9.9219	10		254.000	58	1473.20	106	2692.40	154	3911.60
13/32	.4063	10.3187	11		279.400	59	1498.60	107	2717.80	155	3937.00
27/64	.4219	10.7156	12		304.800	60	1524.00	108	2743.20	156	3962.40
7/16	.4375	11.1125	13		330.200	61	1549.40	109	2768.60	157	3987.80
29/64	.4531	11.5094	14		355.600	62	1574.80	110	2794.00	158	4013.20
15/32	.4688	11.9062	15		381.000	63	1600.20	111	2819.40	159	4038.60
31/64	.4844	12.3031	16		406.400	64	1625.60	112	2844.80	160	4064.00
1/2	.5000	12.7000	17		431.800	65	1651.00	113	2870.20	161	4089.40
33/64	.5156	13.0969	18		457.200	66	1676.40	114	2895.60	162	4114.80
17/32	.5313	13.4937	19		482.600	67	1701.80	115	2921.00	163	4140.20
35/64	.5469	13.8906	20		508.000	68	1727.20	116	2946.40	164	4165.60
9/16	.5625	14.2875	21		533.400	69	1752.60	117	2971.80	165	4191.00
37/64	.5781	14.6844	22		558.800	70	1778.00	118	2997.20	166	4216.40
19/32	.5938	15.0812	23		584.200	71	1803.40	119	3022.60	167	4241.80
39/64	.6094	15.4781	24		609.600	72	1828.80	120	3048.00	168	4267.20
5/8	.6250	15.8750	25		635.000	73	1854.20	121	3073.40	169	4292.60
41/64	.6406	16.2719	26		660.400	74	1879.60	122	3098.80	170	4318.00
21/32	.6563	16.6687	27		685.800	75	1905.00	123	3124.20	171	4343.40
43/64	.6719	17.0656	28		711.200	76	1930.40	124	3149.60	172	4368.80
11/16	.6875	17.4625	29		736.600	77	1955.80	125	3175.00	173	4394.20
45/64	.7031	17.8594	30		762.000	78	1981.20	126	3200.40	174	4419.60
23/32	.7188	18.2562	31		787.400	79	2006.60	127	3225.80	175	4445.00
47/64	.7344	18.6531	32		812.800	80	2032.00	128	3251.20		
3/4	.7500	19.0500	33		838.200	81	2057.40	129	3276.60		

0.001" = .0254 mm   0.001 mm = 0.0004"

APPENDIX D

Available N/C Motion Pictures

Bendix Corporation
Industrial Controls Division
12843 Greenfield Road
Detroit, Michigan 48227

Turning a Profit. 16 mm, 45 min., color/sound.

Use of numerical control equipment in modern industry. Features a lathe machine with Bendix 800 control.

Manufacturing by NC. 16 mm, 25 min., color/sound.

Educational in nature; is intended primarily for viewers with little or no background in the field of numerical control. Film is outdated regarding latest equipment; however, many educational institutions still use it because of the basic information it contains.

Progress in NC. 16 mm, 25 min., color/sound.

Follow-up film to *Manufacturing by NC.* Shows current applications. Outdated regarding latest equipment being used.

Cincinnati Milacron, Inc.
4701 Marburg Avenue
Cincinnati, Ohio 45209
Machine Tool Division

Advanced Systems Capability. 16 mm, 17 min., color/sound.

A product sales film showing four examples of large, specialized machine tool systems designed to machine large parts for earth-moving equipment and aerospace hardware. (When ordering, list alternative showing dates.)

The CIM-Xchanger 25 HC NC Machining Center. 16 mm, 8 min., color/sound.

Shows features and applications of a numerically controlled travelling column, horizontal machining center with 90-tool storage.

10V, 10VC Machining Centers. 16 mm, 15 min., color/sound.

Film shows the features and applications of a 10 HP N/C vertical machining center having a 30-tool storage capability.

10HC Machining Center. 16 mm, 10 min., color/sound.

Features and application of N/C 10 HP travelling column, horizontal machining center with 30-tool storage drum.

The Cinturn N/C Chucking Centers. 16 mm, 18 min., color/sound.

A product film about details, features, and capabilities of the line of N/C chucking centers.

The Cleveland Twist Drill Co.
P.O. Box 6656
Cleveland, Ohio 44101
Use and Care of Twist Drills. 16 mm, 23 min., color/sound.
    Describes twist drill terms and construction, drill pointing and drill point angles for various materials, fundamentals of good drilling practice, and data on speeds and feeds. (List alternative showing dates when ordering.)

Ex-Cell-O Corporation
2855 Coolidge
Troy, Michigan 48084
Applications of Automation. 16 mm, 30 min., BW/sound.
    Automated machine tools in action. Some are standard machines on which work-handling equipment has been applied; others are specially designed to fit into automated production lines. A technical film, yet understandable to students and the general public.

Giddings & Lewis-Bickford Machine Company
A Division of Giddings & Lewis, Inc.
820 Hyland Avenue
Kaukauna, Wisconsin 54130
10-V and 15-V NumeriCenter. 16 mm, 13 1/2 min., color/sound.
    Features the economic values of N/C machining with automatic tool changing. Machines bore, drill, mill, and tap; they have 40-tool storage capacity; they show indexers and multiple station arrangements.

Kearney & Trecker Corporation
11000 Theodore Trecker Way
Milwaukee, Wisconsin 53214
Changing Months to Minutes. 16 mm, 14 min., color/sound.
    Presents the Model II, Milwaukee-Matic tape-controlled machining center in action. Features automatic tool changing, absolute positioning, and a combination of contouring and positioning without the use of computer. Multiple operations performed in milling, drilling, boring, and related functions.
Milwaukee-Matic 180. 16 mm, 13 min., color/sound.
    Presents a chronology of a Value Engineering project, showing the development of a high-performance, low-cost machining center. The result of this project, the Milwaukee-Matic 180 Machining Center, is shown in actual operations, depicting innovative development in machine design and computerized numerical control.
Producing for Profit. 16 mm, 18 min., color/sound.
    Introduces Milwaukee-Matic Series E, tape-controlled machining center with automatic tool changer. Describes how the production approach is applied to job lot machining. The machine performs milling, drilling, boring, tapping, and related operations.

In Search of a Better Way. 16 mm, 18 min., color/sound.
Describes combination of N/C technology and production systems to create a new manufacturing concept — the N/C processing line. Includes process details of a specific system built and shipped in 1977, including N/C milling, specialized turning, and head-changing machines providing multiple-spindle drilling, boring, reaming, and tapping.

The Warner & Swasey Company
Turning Machine Division
5701 Carnegie Avenue
Cleveland, Ohio 44103
New Directions in Turning. 16 mm, 18 min., color/sound.
Details the operation and features of Warner & Swasey's 1-SC, 2-SC, and 3-SC N/C turning machines on bar and chucking work.

The Warner & Swasey Company
Wiedemann Division
211 South Gulph Road
King of Prussia, Pennsylvania 19406
Wiedematic and You . . . It's About Time. 16 mm, 16 min., color.
Action scenes of CNC turret punch press and automatic right angle blade-shearing system. Manufacturing facilities, sales and service, and school for customer training.

White-Sundstrand Machine Tool Co.
A Div. of White Consolidated Ind.
3615 Newburg Road
Belvedere, Illinois 61108
Series 80 Omnimil. 16 mm, 15 min., color/sound.
Illustrates large machining centers of different configurations performing boring, milling, drilling, and tapping operations on several workpieces.

GLOSSARY

TERM AND DEFINITION	EXAMPLE
A AXIS. The axis of circular motion of a machine tool member or slide about the X axis. (Usually called alpha.)	
ABSOLUTE ACCURACY. Accuracy as measured from a reference which must be specified.	
ABSOLUTE READOUT. A display of the true slide position as derived from the position commands within the control system.	
ABSOLUTE SYSTEM. A numerical control system in which all positional dimensions, both input and feedback, are given with respect to a common datum point. The alternative is the incremental system.	

Coordinate Positions

Point	X value	Y value
PT1	2	2
PT2	5	2
PT3	4	5

In an absolute system, all points are relative to (0,0), and the absolute coordinates for each of the required points are programmed with respect to (0,0).

TERM AND DEFINITION	EXAMPLE
ACCANDEC. (Acceleration and deceleration) Acceleration and deceleration in feed rate. It provides smooth starts and stops when operating in N/C and when changing from one feed rate value to another.	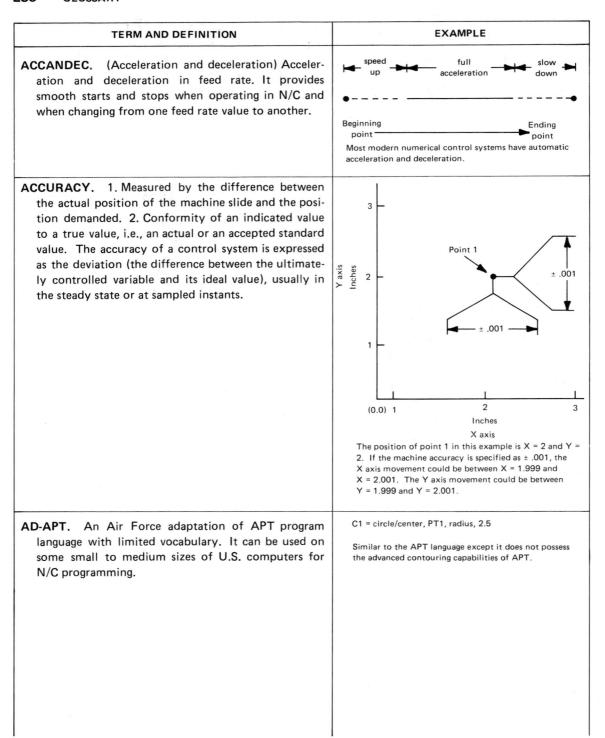
ACCURACY. 1. Measured by the difference between the actual position of the machine slide and the position demanded. 2. Conformity of an indicated value to a true value, i.e., an actual or an accepted standard value. The accuracy of a control system is expressed as the deviation (the difference between the ultimately controlled variable and its ideal value), usually in the steady state or at sampled instants.	
AD-APT. An Air Force adaptation of APT program language with limited vocabulary. It can be used on some small to medium sizes of U.S. computers for N/C programming.	C1 = circle/center, PT1, radius, 2.5 Similar to the APT language except it does not possess the advanced contouring capabilities of APT.

Within the ACCANDEC example:

speed up — full acceleration — slow down

Beginning point → Ending point

Most modern numerical control systems have automatic acceleration and deceleration.

Within the ACCURACY example:

Y axis Inches — Point 1 — ± .001 — ± .001 — (0.0) 1 — 2 — 3 — Inches — X axis

The position of point 1 in this example is X = 2 and Y = 2. If the machine accuracy is specified as ± .001, the X axis movement could be between X = 1.999 and X = 2.001. The Y axis movement could be between Y = 1.999 and Y = 2.001.

TERM AND DEFINITION	EXAMPLE
ADAPTIVE CONTROL. A technique which automatically adjusts feeds and/or speeds to an optimum by sensing cutting conditions and acting upon them.	 Sensors may measure variable factors, e.g. vibration, heat, torque, and deflection. Cutting speeds and feeds may be increased or decreased depending on conditions sensed.
ADDRESS. 1. A symbol indicating the significance of the information immediately following. 2. A means of identifying information or a location in a control system. 3. A number which identifies one location in memory.	
ALPHANUMERIC CODING. A system in which the characters are letters A through Z and numerals 0 through 9.	APT and AD-APT statements use alphanumeric coding, e.g. GOFWD, CT12/PAST, 2, INTOF, L13
ANALOG. 1. Applies to a system which uses electrical voltage magnitudes or ratios to represent physical axis positions. 2. Pertains to information which can have continuously variable values.	
ANALYST. A person skilled in the definition and development of techniques to solve problems.	

APT. (Automatic Programmed Tool) A universal computer-assisted program system for multiaxis contouring programming. APT III provides for five axes of machine tool motion.	Typical APT geometry definition statement:
	    C1 = CIRCLE/XLARGE, L12, XLARGE, L13,
	    RADIUS, 3.5
	
	Typical APT tool motion statement:
	    TLRGT, GORGT/AL3, PAST, AL12

TERM AND DEFINITION	EXAMPLE
ARC CLOCKWISE. An arc generated by the coordinated motion of two axes, in which curvature of the tool path with respect to the workpiece is clockwise, when viewing the plane of motion from the positive direction of the perpendicular axis.	 Activated by G02 preparatory function
ARC COUNTERCLOCKWISE. An arc generated by the coordinated motion of two axes, in which curvature of the tool path with respect to the workpiece is counterclockwise, when viewing the plane of motion from the positive direction of the perpendicular axis.	 Activated by G03 preparatory function
ASCII. (American Standard Code for Information Interchange) A data transmission code which has been established as an American standard by the American Standards Association. It is a code in which seven bits are used to represent each character. Formerly USASCII.	
AUTOMATION. 1. The implementation of processes by automatic means. 2. The investigation, design, development, and application of methods to render processes automatic, self-moving, or self-controlling.	

TERM AND DEFINITION	EXAMPLE
AUTOSPOT. (Automatic System for Positioning of Tools) A computer-assigned program for N/C positioning and straight-cut systems, developed in the U.S. by the IBM Space Guidance Center. It is maintained and taught by IBM.	
AUXILIARY FUNCTION. A programmable function of a machine other than the control of the coordinate movements or cutter.	• Transferring a tool to the select tool position. • Turning coolant ON or OFF. • Starting or stopping the spindle. • Initiating pallet shuttle or movement.
AXIS. A principal direction along which the relative movements of the tool or workpiece occur. There are usually three linear axes, mutually at right angles, designated as X, Y, and Z.	
AXIS INHIBIT. A feature of an N/C unit which enables the operator to withhold command information from a machine tool slide.	
AXIS INTERCHANGE. The capability of inputting the information concerning one axis into the storage of another axis.	
AXIS INVERSION. The reversal of plus and minus values along an axis. This allows the machining of a left-handed part from right-handed programming or vice versa.	

2nd quadrant   1st quadrant
Actual part
(−X, +Y)   (+X, +Y)

3rd quadrant   4th quadrant
(−X, −Y)   (+X, −Y)

TERM AND DEFINITION	EXAMPLE
B (BETA) AXIS. The axis of circular motion of a machine tool member or slide about the Y axis.	
BACKLASH. A relative movement between interacting mechanical parts as a result of looseness.	
BCD. (Binary-coded decimal) A system of number representation in which each decimal digit is represented by a group of binary digits forming a character.	 Numbers and letters are expressed by punched holes across the tape for the code or value desired.

TERM AND DEFINITION	EXAMPLE
BINARY CODE. Based on binary numbers, which are expressed as either 1 or 0, true or false, on or off.	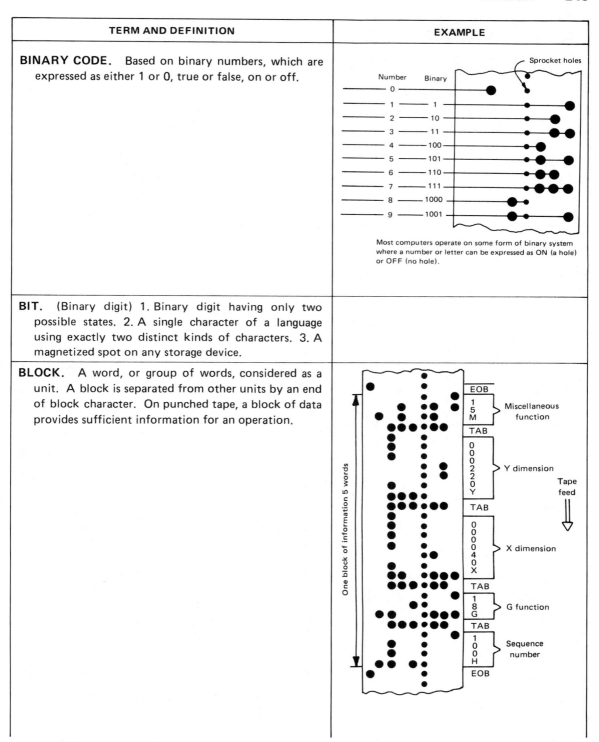
BIT. (Binary digit) 1. Binary digit having only two possible states. 2. A single character of a language using exactly two distinct kinds of characters. 3. A magnetized spot on any storage device.	
BLOCK. A word, or group of words, considered as a unit. A block is separated from other units by an end of block character. On punched tape, a block of data provides sufficient information for an operation.	

TERM AND DEFINITION	EXAMPLE
BLOCK DELETE. Permits selected blocks of tape to be ignored by the control system, at the operator's discretion with permission of the programmer.	This feature allows certain blocks of information to be skipped by programming a slash (/) code in front of the block to be skipped. One lot of parts with holes 1, 2, and 3 are required. On another lot, only holes 1 and 3 are required. The same tape could be used for both lots by activating the block delete switch on the second lot and eliminating hole 2. The (/) code would be in front of the block of information for hole 2.
BUFFER STORAGE. A place for storing information in a control system or computer for planned use. Information from the buffer storage section of a control system can be transferred almost instantly to active storage (that portion of the control system commanding the operation at the particular time). Buffer storage allows a control system to act immediately on stored information rather than wait for the information to be read into the machine from the tape reader.	
BUG. 1. A mistake or malfunction. 2. An integrated circuit (slang).	
BYTE. A sequence of adjacent binary digits usually operated on as a unit and shorter than a computer word.	Eight bits equal one byte. A computer word usually consists of either sixteen or thirty-two bits (two or four bytes).
CAD. Computer aided design	
CAM. (Computer aided manufacturing) The use of computers to assist in phases of manufacturing.	
CAM-I. (Computer Aided Manufacturing International) The outgrowth and replacement organization of the APT Long Range Program.	
CANCEL. A command which will discontinue any canned cycles or sequence commands.	
CANNED CYCLE. A preset sequence of events initiated by a single command. For example, code G84 will perform tap cycle by N/C.	
CARD TO TAPE CONVERTER. A device which converts information directly from punched cards to punched or magnetic tape.	

TERM AND DEFINITION	EXAMPLE
CARTESIAN COORDINATES. A means whereby the position of a point can be defined with reference to a set of axes at right angles to each other.	
C AXIS. Normally the axis of circular motion of a machine tool member or slide about the Z axis.	
CHAD. Pieces of material removed in card or tape operations.	

TERM AND DEFINITION	EXAMPLE
CHANNELS. Paths parallel to the edge of the tape along which information may be stored by the presence or absence of holes or magnetized areas. This term is also known as level or track. The EIA standard one-inch-wide tape has eight channels.	Channels (tracks) 8 7 6 5 4 3 2 1
CHARACTERS. A general term for all symbols, such as alphabetic letters, numerals, and punctuation marks. It is also the coded representation of such symbols.	Character — one line across tape G
CHIP. A single piece of silicon cut from a slice by scribing and breaking. It can contain one or more circuits but is packaged as a unit.	
CIRCULAR INTERPOLATION. 1. Capability of generating up to 360 degrees of arc using only one block of information as defined by EIA. 2. A mode of contouring control which uses the information contained in a single block to produce an arc of a circle.	Y — End point G03 — circular interpolation counterclockwise Start point Center point Center point > I = X coordinate value J = Y coordinate value X

TERM AND DEFINITION	EXAMPLE
CLOSED-LOOP SYSTEM. A system in which the output, or some result of the output, is measured and fed back for comparison with the input. In an N/C system, the output is the position of the table or head; the input is the tape information which ordinarily differs from the output. This difference is measured and results in a machine movement to reduce and eliminate the variance.	
CNC. Computer numerical control	
CODE. A system describing the formation of characters on a tape for representing information, in a language that can be understood and handled by the control system.	
COMMAND. A signal, or series of signals, initiating one step in the execution of a program.	
COMMAND READOUT. A display of the slide position as commanded from the control system.	
CONTINUOUS-PATH OPERATION. An operation in which rate and direction of relative movement of machine members is under continuous numerical control. There is no pause for data reading.	See *contouring control system.*
CONTOURING CONTROL SYSTEM. An N/C system for controlling a machine (e.g. milling, drafting) in a path resulting from the coordinated, simultaneous motion of two or more axes.	

TERM AND DEFINITION	EXAMPLE
COORDINATE DIMENSIONING. A system of dimensioning based on a common starting point.	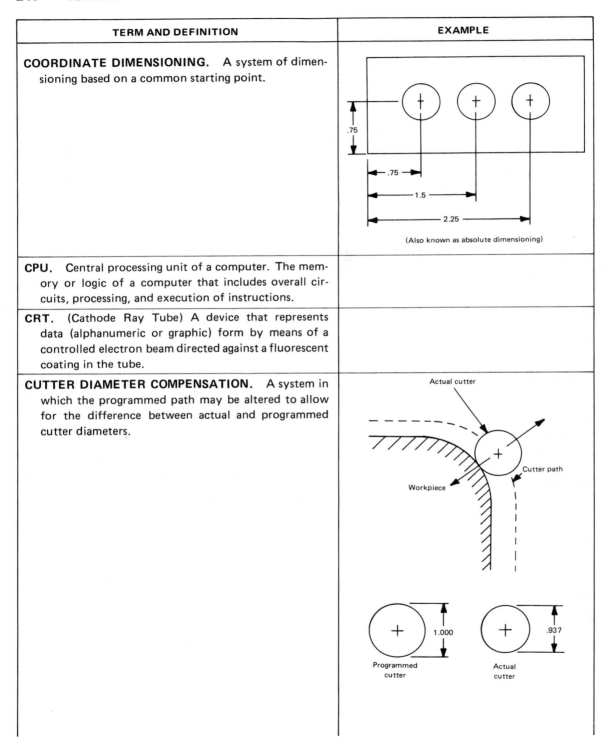
CPU. Central processing unit of a computer. The memory or logic of a computer that includes overall circuits, processing, and execution of instructions.	
CRT. (Cathode Ray Tube) A device that represents data (alphanumeric or graphic) form by means of a controlled electron beam directed against a fluorescent coating in the tube.	
CUTTER DIAMETER COMPENSATION. A system in which the programmed path may be altered to allow for the difference between actual and programmed cutter diameters.	

TERM AND DEFINITION	EXAMPLE
CUTTER OFFSET. The distance from the part surface to the axial center of a cutter.	Offset
CUTTER PATH. The path defined by the center of a cutter.	
CYCLE. 1. A sequence of operations that is repeated regularly. 2. The time it takes for one such sequence to occur.	
DATA. A representation of information in the form of words, symbols, numbers, letters, characters, digits, etc.	
DEBUG. 1. To detect, locate, and remove mistakes from a program. 2. Troubleshoot.	
DECIMAL CODE. A code in which each allowable position has one of ten possible states. (The conventional decimal number system is a decimal code.)	
DELETE CHARACTER. A character used primarily to obliterate any erroneous or unwanted characters on punched tape. The delete character consists of perforations in all punching positions.	
DIAGNOSTIC TEST. The running of a machine program or routine to discover a failure or potential failure of a machine element and to determine its location.	
DIGIT. A character in any numbering system.	

TERM AND DEFINITION	EXAMPLE
DIGITAL. 1. Refers to discrete states of a signal (on or off). A combination of these make up a specific value. 2. Relating to data in the form of digits.	
DISPLAY. A visual representation of data.	
DOCUMENTATION. Manuals and other printed materials (tables, magnetic tape, listing, diagrams) which provide information for use and maintenance of a manufactured product, both hardware and software.	
DWELL TIME. A timed delay of programmed or established duration, not cyclic or sequential. It is not an interlock or hold time.	
EDIT. To modify the form of data.	
EIA STANDARD CODE. A standard code for positioning, straight-cut, and contouring control systems proposed by the U.S. EIA in their Standard RS-244. Eight-track paper (one-inch wide) has been accepted by the American Standards Association as an American standard for numerical control.	
END OF BLOCK CHARACTER. 1. A character indicating the end of a block of tape information. Used to stop the tape reader after a block has been read. 2. The typewriter function of the carriage return when preparing machine control tapes.	EOB character
END OF PROGRAM. A miscellaneous function (M02) indicating the completion of a workpiece. Stops spindle, coolant, and feed after completion of all commands in the block. Used to reset control and/or machine.	
END OF TAPE. A miscellaneous function (M30) which stops spindle, coolant, and feed after completion of all commands in the block. Used to reset control and/or machine.	
END POINT. The extremities of a span.	
ERROR SIGNAL. Indication of a difference between the output and input signals in a servo system.	

TERM AND DEFINITION	EXAMPLE
EXECUTIVE PROGRAM. A series of programming instructions enabling a dedicated minicomputer to produce a specific output control. For example, it is the executive program in a CNC unit that enables the control to think like a lathe or machining center.	
FEED. The programmed or manually established rate of movement of the cutting tool into the workpiece for the required machining operation.	
FEEDBACK. The transmission of a signal from a late to an earlier stage in a system. In a closed-loop N/C system, a signal of the machine slide position is fed back and compared with the input signal, which specifies the demanded position. These two signals are compared and generate an error signal if a difference exists.	
FEED FUNCTION. The relative motion between the tool or instrument and the work due to motion of the programmed axis.	
FEED RATE (CODE WORD). A multiple-character code containing the letter F followed by digits. It determines the machine slide rate of feed.	
FEED RATE DIVIDER. A feature of some machine control units that gives the capability of dividing the programmed feed rate by a selected amount as provided for in the machine control unit.	
FEED RATE MULTIPLIER. A feature of some machine control units that gives the capability of multiplying the programmed feed rate by a selected amount as provided for in the machine control unit.	
FEED RATE OVERRIDE. A variable manual control function directing the control system to reduce the programmed feed rate.	Feedrate override is a percentage function to reduce the programmed feed rate. If the programmed feed rate was 30 inches per minute and the operator wanted 15 inches per minute, the feedrate override dial would be set at 50 percent.
FIXED BLOCK FORMAT. A format in which the number and sequence of words and characters appearing in successive blocks is constant.	
FIXED CYCLE. See canned cycle.	
FIXED SEQUENTIAL FORMAT. A means of identifying a word by its location in a block of information. Words must be presented in a specific order, and all possible words preceding the last desired word must be present in the block.	

TERM AND DEFINITION	EXAMPLE
FLOATING ZERO. A characteristic of a machine control unit permitting the zero reference point on an axis to be established readily at any point in the travel.	 Machine table The part or workpiece may be moved to *any* location on the machine table and zero may be established at that point.
FORMAT (TAPE). The general order in which information appears on the input media, such as the location of holes on a punched tape or the magnetized areas on a magnetic tape.	
FULL RANGE FLOATING ZERO. A characteristic of a numerical machine tool control permitting the zero point on an axis to be shifted readily over a specified range. The control retains information on the location of permanent zero.	 Machine table The part or workpiece may be shifted to any position on the machine table, but the actual position of permanent zero remains constant.

TERM AND DEFINITION	EXAMPLE
GAGE HEIGHT. A predetermined partial retraction point along the Z axis to which the cutter retreats from time to time to allow safe XY table travel.	 Gage height, usually .100 to .125, is a set distance established in the control or set by the operator. Gage height allows the tool, while advancing in rapid traverse, to stop at the established distance (gage height) and begin feed motion. Without gage height, the tool would rapid into the part causing tool damage or breakage and potential operator injury.
G CODE. A word addressed by the letter G and followed by a numerical code defining preparatory functions or cycle types in a numerical control system.	 G81 — Drill Cycle
GENERAL PROCESSOR. 1. A computer program for converting geometric input data into cutter path data required by an N/C machine. 2. A fixed software program designed for a specific logical manipulation of data.	

TERM AND DEFINITION	EXAMPLE
HARD COPY. A readable form of data output on paper.	
HARDWARE. The component parts used to build a computer or control system, e.g. integrated circuits, diodes, transistors.	
HARD-WIRED. Having logic circuits interconnected on a backplane to give a fixed pattern of events.	
HIGH-SPEED READER. A reading device which can be connected to a computer or control so as to operate on line without seriously holding up the computer or control.	
HOLLERITH. Pertaining to a particular type of code or punched card, using twelve rows per column and usually eighty columns per card.	
INCREMENTAL DIMENSIONING. The method of expressing a dimension with respect to the preceding point in a sequence of points.	
INCREMENTAL SYSTEM. A control system in which each coordinate or positional dimension, both input and feedback, is taken from the last position rather than from a common datum point, as in the absolute system.	 Coordinate positions Point X value Y value PT1 2 5 PT2 3 -2 PT3 -2 -2 In an incremental system, all points are expressed relative to the preceding point.
INDEX TABLE. A multiple-character code containing the letter B followed by digits. This code determines the position of the rotary index table in degrees.	See B (Beta) axis.

TERM AND DEFINITION	EXAMPLE
INHIBIT. To prevent an action or acceptance of data by applying an appropriate signal to the appropriate input.	
INPUT. Transfer of external information into the control system.	
INPUT MEDIA. 1. The form of input such as punched cards and tape or magnetic tape. 2. The device used to input information.	
INTERCHANGEABLE VARIABLE BLOCK FORMAT. A programming arrangement consisting of a combination of the word address and tab sequential formats to provide greater compatibility in programming. Words are interchangeable within the block. Length of block varies since words may be omitted.	This is one of the most sophisticated tape formats in use today. See *block*.
INTERCHANGE STATION. The position where a tool of an automatic tool changing machine awaits automatic transfer to either the spindle or the appropriate coded drum station.	
INTERMEDIATE TRANSFER ARM. The mechanical device in automatic tool changing that grips and removes a programmed tool from the coded drum station and places it into the interchange station, where it awaits transfer to the machine spindle. This device then automatically grips and removes the used tool from the interchange station and returns it to the appropriate coded drum station.	
INTERPOLATION. 1. The insertion of intermediate information based on an assumed order or computation. 2. A function of a control whereby data points are generated between given coordinate positions.	
INTERPOLATOR. A device which is part of a numerical control system and performs interpolation.	
ISO. International Organization for Standardization.	
JOG. A control function which momentarily operates a drive to the machine.	
LEADING ZEROES. Redundant zeroes to the left of a number.	Leading zeroes $X + \overset{\frown}{00}62500$
LEADING ZERO SUPPRESSION. See zero suppression.	

TERM AND DEFINITION	EXAMPLE
LETTER ADDRESS. The method by which information is directed to different parts of the system. All information must be preceded by its proper letter address, e.g., X, Y, Z, M.	 X and G address An identifying letter inserted in front of each word.
LINEAR INTERPOLATION. A function of a control whereby data points are generated between given coordinate positions to allow simultaneous movement of two or more axes of motion in a linear (straight) path.	 The control system moves X and Y axes proportionately to arrive at the destination point.
LOOP TAPE. A short piece of tape, with joined ends which contains a complete program or operation.	
MACHINING CENTER. Machine tools, usually numerically controlled, capable of automatically drilling, reaming, tapping, milling, and boring multiple faces of a part. They are often equipped with a system for automatically changing cutting tools.	

TERM AND DEFINITION	EXAMPLE

MACRO. A group of instructions which can be stored and recalled as a group to solve a recurring problem.	An APT macro could be as follows:
	
	DRILL1 = MACRO/X, Y, Z, Z1, FR, RR
	   GOTO/POINT, X, Y, Z, RR
	   GODLTA/−Z1, FR
	   GODLTA/+Z1, RR
	TERMAC
	
	X, Y, Z, Z1, FR, and RR would be variables which would have values assigned when the macro is called into action. The variables would be as follows:
	    X = X position
	    Y = Y position
	    Z = Z position (above work surface)
	    Z1 = Z feed distance
	    FR = feed rate
	    RR = rapid rate
	
	The call statement could be:
	    CALL/DRILL1, X = 2, Y = 4, Z = .100, Z1 = 1.25,
	    FR = 2, RR = 200

MAGIC-THREE CODING. A feed rate code that uses three digits of data in the F word. The first digit defines the power of ten multiplier. It determines the positioning of the floating decimal point. The last two digits are the most significant digits of the desired feed rate.	To program a feed rate of 12 inches per minute in magic-three coding: 1) count the number of decimal places to the left of the decimal. 12 = 2 2) Add magic "3" to the number of counted decimal places. (3 + 2 = 5) 3) write the F word address, the added digit, and the first two digits of the actual feed rate to be programmed. (F512) 4) F512 would be the magic "3" coded feed rate. This method of feed rate coding is now almost obsolete.
MAGNETIC TAPE. A tape made of plastic and coated with magnetic material. It stores information by selective polarization of portions of the surface.	
MANUAL DATA INPUT. A mode or control that enables an operator to insert data into the control system. This data is identical to information that could be inserted by tape.	
MANUAL PART PROGRAMMING. The preparation of a manuscript in machine control language and format to define a sequence of commands for use on an N/C machine.	Manual, or hand, programming is programming the actual codes, X and Y positions, functions, etc. as they are punched in the N/C tape. H001 G81 X+37500 Y+52500 W01
MANUSCRIPT. A written or printed copy, in symbolic form, containing the same data as that punched on cards or tape or retained in a memory unit.	

TERM AND DEFINITION	EXAMPLE
MEMORY. An organized collection of storage elements, e.g., disc, drum, ferrite cores, into which a unit of information consisting of a binary digit can be stored and from which it can later be retrieved.	A computer with a 64,000-word capacity is said to have a memory of 64 K.
MIRROR IMAGE. See axis inversion.	
MODAL. Information that is retained by the system until new information is obtained and replaces it.	
MODULE. An interchangeable plug-in item containing components.	
N/C. (Numerical control) The technique of controlling a machine or process by using command instructions in coded numerical form.	
NULL. 1. Pertaining to no deflection from a center or end position. 2. Pertaining to a balanced or zero output from a device.	
NUMERICAL CONTROL SYSTEM. A system in which programmed numerical values are directly inserted, stored on some form of input medium, and automatically read and decoded to cause a corresponding movement in a machine or process.	
OFFSET. A displacement in the axial direction of the tool which is the difference between the actual tool length and the programmed tool length.	
OPEN-LOOP SYSTEM. A control system that has no means of comparing the output with the input for control purposes. No feedback.	Instructions Machine table Lead screw Drive motor Control
OPTIMIZE. To rearrange the instructions or data in storage so that a minimum number of transfers are required in the running of a program. To obtain maximum accuracy and minimum part production time by manipulation of the program.	

TERM AND DEFINITION	EXAMPLE
OPTIONAL STOP. A miscellaneous function (M01) command similar to Program Stop except the control ignores the command unless the operator has previously pushed a button to validate the command.	
OVERSHOOT. A term applied when the motion exceeds the target value. The amount of overshoot depends on the feed rate, the acceleration of the slide unit, or the angular change in direction.	Workpiece · Cutter path · Cutter · Overshoot
PARABOLA. A plane curve generated by a point moving so that its distance from a fixed second point is equal to its distance from a fixed line.	
PARABOLIC INTERPOLATION. Control of cutter path by interpolation between three fixed points by assuming the intermediate points are on a parabola.	
PARITY CHECK. 1. A hole punched in one of the tape channels whenever the total number of holes is even, to obtain an odd number, or vice versa depending on whether the check is even or odd. 2. A check that tests whether the number of ones (or zeroes) in any array of binary digits is odd or even.	1 — Already odd number of holes 5 — Even number of holes gets automatic punch in Track 5 to yield odd number of holes in that row. Track 5 RS-244-A (EIA or BCD)

TERM AND DEFINITION	EXAMPLE
PART PROGRAM. A specific and complete set of data and instructions written in source languages for computer processing or in machine language for manual programming to manufacture a part on an N/C machine.	
PART PROGRAMMER. A person who prepares the planned sequence of events for the operation of a numerically controlled machine tool.	
PERFORATED TAPE. A tape on which a pattern of holes or cuts is used to represent data.	
PLOTTER. A device which will draw a plot or trace from coded N/C data input.	
POINT-TO-POINT CONTROL SYSTEM. A numerical control system in which controlled motion is required only to reach a given end point, with no path control during the transition from one end point to the next.	
POSITIONING/CONTOURING. A type of numerical control system that has the capability of contouring, without buffer storage, in two axes and positioning in a third axis for such operations as drilling, tapping, and boring.	
POSITIONING SYSTEM. See point-to-point control system.	
POSITION READOUT. A display of absolute slide position as derived from a position feedback device (transducer) normally attached to the lead screw of the machine. See command readout.	

TERM AND DEFINITION	EXAMPLE
POSTPROCESSOR. The part of the software which converts the cutter path coordinate data into a form which the machine control can interpret correctly. The cutter path coordinate data is obtained from the general processor and all other programming instructions and specifications for the particular machine and control.	
PREPARATORY FUNCTION. An N/C command on the input tape changing the mode of operation of the control. (Generally noted at the beginning of a block by the letter G plus two digits.)	Some preparatory functions are: G84 — tap cycle G01 — linear interpolation G82 — dwell cycle G02 — circular interpolation — clockwise G03 — circular interpolation — counter clockwise See *G code.*
PROGRAM. A sequence of steps to be executed by a control or a computer to perform a given function.	
PROGRAMMED DWELL. The capability of commanding delays in program execution for a programmable length of time.	
PROGRAMMER (PART PROGRAMMER). A person who prepares the planned sequence of events for the operation of a numerically controlled machine tool. The programmer's principal tool is the manuscript on which the instructions are recorded.	Manual part programming instructions: H001 G81 X+123750 Y+62500 W01 N002 X+105000 N003 Y+51250 M06 Computer part programming instructions: TLRGT, GORGT/HL3, TANTO, C1 GOFWD/C1, TANTO, HL2 GOFWD/HL2, PAST, VL2
PROGRAM STOP. A miscellaneous function (M00) command to stop the spindle, coolant, and feed after completion of the dimensional move commanded in the block. To continue with the remainder of the program, the operator must push a button.	

TERM AND DEFINITION	EXAMPLE
QUADRANT. Any of the four parts into which a plane is divided by rectangular coordinate axes in that plane.	
RANDOM. Not necessarily in a logical order of arrangement according to usage, but having the ability to select from any location and in any order from the storage system.	
RAPID. Positioning the cutter and workpiece into close proximity with one another at a high rate of travel speed, usually 150 to 400 inches per minute (IPM) before the cut is started.	
READER. A pneumatic, photoelectric, or mechanical device used to sense bits of information on punched cards, punched tape, or magnetic tape.	
REGISTER. An internal array of hardware binary circuits for temporary storage of information.	
REPEATABILITY. Closeness of, or agreement in, repeated measurements of the same characteristics by the same method, using the same conditions.	
RESET. To return a register or storage location to zero or to a specified initial condition.	
ROW (TAPE). A path perpendicular to the edge of the tape along which information may be stored by the presence or absence of holes or magnetized areas. A character would be represented by a combination of holes.	

For the QUADRANT example:

+Y

2nd quadrant (−X, +Y)   1st quadrant (+X, +Y)

−X ———————————— +X

3rd quadrant (−X, −Y)   4th quadrant (+X, −Y)

−Y

For the ROW (TAPE) example: Row

TERM AND DEFINITION	EXAMPLE
SEQUENCE NUMBER (CODE WORD). A series of numerals programmed on a tape or card and sometimes displayed as a readout; normally used as a data location reference or for card sequencing.	
SEQUENCE READOUT. A display of the number of the block of tape being read by the tape reader.	
SEQUENTIAL. Arranged in some predetermined logical order.	
SIGNIFICANT DIGIT. A digit that must be kept to preserve a specific accuracy or precision.	Significant digits $X + 00\overset{\frown}{525}00$ Insignificant digits

SLOW-DOWN SPAN. A span of information having the necessary length to allow the machine to decelerate from the initial feed rate to the maximum allowable cornering feed rate that maintains the specified tolerance.	
	
	Postprocessor output
	
	G01 X+42500 Y+100000 F200
	      X+19750 F175
	      X+17215 F140
	      X+13750
	                 Y+68750 F200

TERM AND DEFINITION	EXAMPLE
SOFTWARE. Instructional literature and computer programs used to aid in part programming, operating, and maintaining the machining center.	Examples of software programs are: APT FORTRAN COBOL RPG
SPAN. A certain distance or section of a program designated by two end points for linear interpolation; a beginning point, a center point, and an ending point for circular interpolation; and two end points and a diameter point for parabolic interpolation.	 One linear interpolation span
SPINDLE SPEED (CODE WORD). A multiple-character code containing the letter S followed by digits. This code determines the RPM of the cutting spindle of the machine.	
STORAGE. A device into which information can be introduced, held, and then extracted at a later time.	
TAB. A nonprinting spacing action on tape preparation equipment. A tab code is used to separate words or groups of characters in the tab sequential format. The spacing action sets typewritten information on a manuscript into tabular form.	 Tab code
TAB SEQUENTIAL FORMAT. Means of identifying a word by the number of tab characters preceding the word in a block. The first character of each word is a tab character. Words must be presented in a specific order, but all characters in a word, except the tab character, may be omitted when the command represented by that word is not desired.	*005 *1 *07000 *16000 *1 Seq. No Prep. function X dimension Y dimension Miscellaneous function * = Tab The tab sequential format is, for the most part, obsolete.

TERM AND DEFINITION	EXAMPLE
TAPE. A magnetic or perforated paper medium for storing information.	
TAPE LAGGER. The trailing end portion of a tape. **TAPE LEADER.** The front or lead portion of a tape.	 Reel tapes should have a leader and lagger of approximately three feet with just sprocket holes for tape loading and threading purposes.
TOOL FUNCTION. A tape command identifying a tool and calling for its selection. The address is normally a T word.	T06 would be a tape command calling for the tool assigned to spindle or pocket 6 to be put in the spindle.
TOOL LENGTH COMPENSATION. A manual input, by means of selector switches, to eliminate the need for preset tooling; allows the programmer to program all tools as if they are of equal length.	
TOOL OFFSET. 1. A correction for tool position parallel to a controlled axis. 2. The ability to reset tool position manually to compensate for tool wear, finish cuts, and tool exchange.	 Tool offsets are used as final adjustments to increase or decrease depths due to cutting forces and tool deflection. In this case, a tool offset could be used to increase the drill depth from depth-1 to depth-2.
TRAILING ZERO SUPPRESSION. See zero suppression.	

TERM AND DEFINITION	EXAMPLE
TURN KEY SYSTEM. A term applied to an agreement whereby a supplier will install an N/C or computer system so that he has total responsibility for building, installing, and testing the system.	
USASCII. United States of America Standard Code for Information Interchange. See ASCII.	
VARIABLE BLOCK FORMAT (TAPE). A format which allows the quantity of words in successive blocks to vary.	Same as word address. Variable block means the length of the blocks can vary depending on what information needs to be conveyed in a given block. See *block*.
VECTOR. A quantity that has magnitude, direction, and sense; is represented by a directed line segment whose length represents the magnitude and whose orientation in space represents the direction.	
VECTOR FEED RATE. The feed rate at which a cutter or tool moves with respect to the work surface. The individual slides may move slower or faster than the programmed rate, but the resultant movement is equal to the programmed rate.	
WORD. An ordered set of characters which is the normal unit in which information may be stored, transmitted, or operated upon.	
WORD ADDRESS FORMAT. The specific arrangement of addressing each word in a block of information by one or more alphabetical characters which identify the meaning of the word.	See *address* and *block*.
WORD LENGTH. The number of bits or characters in a word.	See *word*.
X AXIS. Axis of motion that is always horizontal and parallel to the workholding surface.	
Y AXIS. Axis of motion that is perpendicular to both the X and Z axes.	

TERM AND DEFINITION	EXAMPLE
Z AXIS. Axis of motion that is always parallel to the principal spindle of the machine.	
ZERO OFFSET. A characteristic of a numerical machine tool control permitting the zero point on an axis to be shifted readily over a specified range. The control retains information on the location of the permanent zero.	See *full range floating zero* and *floating zero*.
ZERO SHIFT. A characteristic of a numerical machine tool control permitting the zero point on an axis to be shifted readily over a specified range. (The control does *not* retain information on the location of the permanent zero.)	See *floating zero*. Consult chapter 4 for additional details.
ZERO SUPPRESSION. Leading zero suppression: the elimination of insignificant leading zeroes to the left of significant digits usually before printing. Trailing zero suppression: the elimination of insignificant trailing zeroes to the right of significant digits usually before printing.	Leading zero suppression X + 0043500 Insignificant digits Could be written as: X + 43500 Trailing zero suppression X + 0043500 Insignificant digits Could be written as: X + 00435

INDEX